GRADUATE STUDIES
TEXAS TECH UNIVERSITY

Reactivity Indices for Biomolecules

Chen-An Chin and Pill-Soon Song

No. 24 October 1981

TEXAS TECH UNIVERSITY

Lauro F. Cavazos, President

Graduate Studies No. 24
175 pp.
9 October 1981
Paper, $20.00
Cloth, $33.00

Graduate Studies are numbered serially, paged separately, and published on an irregular basis under the auspices of the Dean of the Graduate School and Director of Academic Publications, and in cooperation with the International Center for Arid and Semi-Arid Land Studies. The preferred abbreviation for citing this series is Grad. Studies, Texas Tech Univ. Copies can be purchased on standing order or separately by title from the Texas Tech Press Sales Office, Texas Tech University Library, Lubbock, Texas 79409, U.S.A. Orders from individuals should be accompanied by remittance, and all payments should be made in U.S. currency, with check, money order, or bank draft drawn on a U.S. bank. Prices include normal handling and postage when mailed within the United States (add $2.00 per title for foreign postage; Texas residents must pay a 5 per cent sales tax on total purchase price). Institutions interested in exchanging publications should address the Exchange Librarian at Texas Tech University.

ISSN 0082-3198
ISBN 0-89672-092-6 (paper)
ISBN 0-89672-093-4 (cloth)

Texas Tech Press, Lubbock, Texas

1981

CONTENTS

Reactivity Indices for Biomolecules

Chen-An Chin and Pill-Soon Song

The π-electron reactivity indices of biomolecules presented in this volume have been calculated in accordance with the Hückel Molecular Orbital theory. The calculated reactivity indices are:

1. Electron Density (P_{rr})
2. Superdelocalizability for Nucleophilic Attack (SDN_r)
3. Frontier Orbital Density (FOD_r)
4. Superdelocalizability for Radical Attack (SDR_r)
5. Frontier Radical Density (FRD_r)
6. Superdelocalizability for Electrophilic Attack (SDE_r)
7. Frontier Electron Density (FED_r)
8. Atom-Atom Polarizability (π_{rr})

The subscript "r" refers to the rth atom. The reactivity indices for each atom are shown in separate Tables for each molecular structure, and the atoms in each molecule are numbered arbitrarily.

The reactivity indices for Frontier Density and Superdelocalizability are derived from Fukui's Frontier Electron Theory.[1,2,3] The Atom-Atom Polarizability index is derived from Coulson and Longuet-Higgins[4] on the basis of perturbation theory. Definitions of these reactivity indices are shown in Appendix A. Coulomb (α) and Resonance (β) integral parameters of Pullman (designated P)[5], Streitwieser (designated S)[6] and Fukui (designated F) are shown in Appendix B. Methyl groups are treated with the group orbital (hyperconjugation) model (shown as $-C\equiv H_3$)[7] except where the inductive model (I)[8,9] is employed (-0.5 for the Coulomb parameter for the carbon atom carrying a methyl group). Saturated side chains and rings (for example, ionone ring of carotenoids and retinol) have also been treated by the same group orbital model (shown as $-C=H_2$ or $-C\equiv H_3$). For example, the alanyl side chain of tyrosine is treated by the Methyl group orbital. The Mg^{++} ions in Chlorophylls and the Fe^{++} ion in Heme have been approximated as point charges in evaluating an appropriate set of Coulomb and resonance integrals. For other details of the calculation and the use of π-electron reactivity indices, the readers may refer to our previous publications.[10,11]

ACKNOWLEDGEMENTS

The authors acknowledge the support of these calculations by the Robert A. Welch Foundation (Grant D-182). The authors express their thanks to Mrs. Laurie Robbins for her skillful drawing of the structural diagrams, and to Dr. Yukio Yamaguchi for his valuable assistance in the computer programming of the calculations. Chin also expresses his deep appreciation to his daughter, Weni, and his son, Yang, for their help with the painstaking job of typing the calculated reactivity indices.

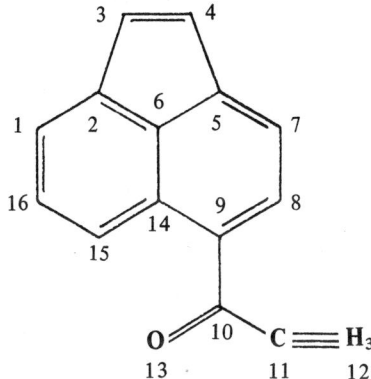

TABLE 1(P).—5-Acetoacenaphthylene.

Atom	P_{rr}	SDN_r	FOD_r	SDR_r	FRD_r	SDE_r	FED_r	π_{rr}
1	0.906	1.874	0.260	1.315	0.174	0.756	0.089	0.449
2	1.043	0.931	0.091	0.890	0.089	0.849	0.087	0.349
3	1.049	1.474	0.212	1.309	0.353	1.144	0.493	0.484
4	1.069	1.275	0.160	1.234	0.359	1.192	0.558	0.472
5	1.020	1.136	0.141	0.970	0.095	0.805	0.049	0.358
6	1.094	0.612	0.000	0.830	0.176	1.047	0.353	0.353
7	0.915	1.802	0.245	1.284	0.166	0.766	0.087	0.447
8	0.968	1.140	0.073	0.974	0.037	0.809	0.001	0.418
9	0.963	1.836	0.305	1.318	0.202	0.800	0.099	0.416
10	0.797	1.015	0.053	0.643	0.027	0.271	0.002	0.212
11	0.594	0.491	0.000	0.491	0.000	0.491	0.000	0.242
12	1.029	0.535	0.007	0.540	0.003	0.544	0.000	0.259
13	1.279	0.972	0.107	0.772	0.064	0.572	0.021	0.260
14	0.982	0.771	0.006	0.730	0.012	0.688	0.018	0.336
15	0.930	1.952	0.303	1.393	0.218	0.833	0.133	0.467
16	1.002	0.927	0.038	0.886	0.024	0.844	0.011	0.400

TABLE 1(S).—5-Acetoacenaphthylene.

Atom	P_{rr}	SDN_r	FOD_r	SDR_r	FRD_r	SDE_r	FED_r	π_{rr}
1	0.906	1.871	0.261	1.314	0.175	0.756	0.089	0.449
2	1.043	0.930	0.092	0.890	0.089	0.849	0.087	0.349
3	1.048	1.470	0.213	1.307	0.353	1.144	0.493	0.484
4	1.069	1.274	0.161	1.233	0.360	1.192	0.559	0.472
5	1.020	1.131	0.141	0.968	0.095	0.805	0.049	0.358
6	1.094	0.612	0.000	0.830	0.176	1.047	0.353	0.353
7	0.915	1.800	0.246	1.283	0.167	0.766	0.087	0.447
8	0.968	1.135	0.073	0.972	0.037	0.809	0.001	0.417
9	0.964	1.834	0.307	1.317	0.203	0.800	0.099	0.416
10	0.794	1.004	0.052	0.637	0.027	0.270	0.002	0.211
11	1.081	0.277	0.000	0.332	0.000	0.386	0.000	0.161
12	0.902	0.384	0.004	0.385	0.002	0.331	0.000	0.170
13	1.282	0.942	0.103	0.756	0.062	0.570	0.021	0.257
14	0.982	0.772	0.006	0.729	0.012	0.688	0.018	0.336
15	0.930	1.942	0.305	1.391	0.219	0.833	0.133	0.468
16	1.002	0.926	0.038	0.885	0.024	0.844	0.011	0.400

TABLE 2.—Adenine.

Atom	P_{π}	SDN_r	FOD_r	SDR_r	FRD_r	SDE_r	FED_r	π_{rr}
1	1.180	0.184	0.108	1.064	0.285	1.944	0.462	0.160
2	0.867	0.966	0.464	0.817	0.308	0.669	0.151	0.346
3	1.282	0.584	0.003	0.890	0.069	1.196	0.135	0.371
4	0.898	0.998	0.369	0.888	0.248	0.779	0.128	0.401
5	1.267	0.643	0.341	0.949	0.317	1.255	0.292	0.387
6	0.063	0.762	0.017	0.789	0.057	0.815	0.097	0.355
7	1.593	0.345	0.068	0.790	0.075	1.235	0.082	0.246
8	0.928	0.981	0.454	0.984	0.340	0.986	0.226	0.444
9	1.305	0.541	0.120	0.902	0.180	1.263	0.239	0.359
10	1.087	0.576	0.055	0.807	0.121	1.037	0.188	0.339

TABLE 3(P).—Aflatoxin *B*.

Atom	P_{rr}	SDN_r	FOD_r	SDR_r	FRD_r	SDE_r	FED_r	π_{rr}
1	1.917	0.088	0.025	0.570	0.057	1.053	0.088	0.051
2	0.918	0.965	0.200	0.879	0.209	0.794	0.218	0.376
3	1.127	0.640	0.000	0.922	0.000	1.205	0.000	0.405
4	0.909	1.006	0.197	0.921	0.202	0.835	0.209	0.390
5	1.914	0.093	0.045	0.576	0.055	1.058	0.084	0.054
6	1.059	0.480	0.000	0.568	0.022	0.656	0.043	0.263
7	0.952	0.487	0.000	0.487	0.002	0.487	0.004	0.239
8	1.089	0.630	0.007	0.936	0.139	1.243	0.273	0.391
9	0.918	0.986	0.165	0.906	0.085	0.816	0.005	0.389
10	1.099	0.558	0.011	0.788	0.154	1.018	0.296	0.333
11	0.809	1.484	0.626	1.040	0.335	0.595	0.043	0.393
12	0.934	0.487	0.006	0.487	0.003	0.487	0.001	0.239
13	1.029	0.581	0.084	0.577	0.045	0.572	0.007	0.266
14	1.834	0.165	0.046	0.560	0.023	0.955	0.000	0.082
15	0.807	0.757	0.038	0.521	0.024	0.286	0.010	0.206
16	1.149	0.751	0.250	1.042	0.372	1.333	0.495	0.407
17	0.808	0.866	0.109	0.576	0.060	0.287	0.010	0.214
18	0.954	0.492	0.001	0.491	0.001	0.492	0.000	0.242
19	1.030	0.518	0.015	0.532	0.008	0.547	0.002	0.259
20	1.399	0.441	0.051	0.622	0.078	0.801	0.105	0.235
21	1.323	0.628	0.146	0.657	0.126	0.686	0.107	0.256

TABLE 3(S).—Aflatoxin B.

Atom	P_{rr}	SDN_r	FOD_r	SDR_r	FRD_r	SDE_r	FED_r	π_{rr}
1	1.918	0.085	0.027	0.570	0.056	1.055	0.089	0.051
2	0.923	0.948	0.214	0.874	0.218	0.800	0.221	0.375
3	1.127	0.640	0.000	0.923	0.000	1.206	0.000	0.405
4	0.912	0.991	0.204	0.917	0.208	0.843	0.213	0.390
5	1.915	0.091	0.025	0.576	0.056	1.061	0.086	0.053
6	0.920	0.355	0.001	0.275	0.012	0.400	0.033	0.173
7	1.080	0.276	0.000	0.333	0.001	0.389	0.003	0.161
8	1.084	0.637	0.008	0.947	0.143	1.257	0.278	0.396
9	0.923	0.969	0.178	0.896	0.091	0.823	0.004	0.388
10	1.101	0.555	0.016	0.787	0.158	1.019	0.299	0.333
11	0.801	1.465	0.642	1.030	0.343	0.594	0.044	0.394
12	1.080	0.276	0.000	0.331	0.000	0.385	0.000	0.161
13	0.902	0.412	0.046	0.381	0.025	0.350	0.004	0.173
14	1.834	0.161	0.048	0.559	0.023	0.957	0.000	0.091
15	0.808	0.754	0.039	0.520	0.025	0.286	0.010	0.206
16	1.158	0.717	0.247	1.030	0.373	1.342	0.500	0.400
17	0.805	0.855	0.107	0.570	0.059	0.286	0.010	0.213
18	1.081	0.277	0.000	0.332	0.000	0.387	0.000	0.161
19	0.902	0.373	0.008	0.353	0.004	0.333	0.001	0.169
20	1.400	0.437	0.051	0.619	0.078	0.802	0.105	0.234
21	1.327	0.577	0.139	0.642	0.124	0.685	0.109	0.252

TABLE 4(P).—Aflatoxin G.

Atom	P_{rr}	SDN_r	FOD_r	SDR_r	FRD_r	SDE_r	FED_r	π_{rr}
1	1.717	0.086	0.026	0.570	0.057	1.054	0.088	0.051
2	0.919	0.957	0.208	0.877	0.214	0.797	0.220	0.376
3	1.127	0.640	0.000	0.922	0.000	1.205	0.000	0.405
4	0.910	0.998	0.201	0.918	0.205	0.838	0.208	0.389
5	1.915	0.092	0.025	0.576	0.054	1.059	0.084	0.054
6	1.059	0.480	0.001	0.568	0.021	0.656	0.042	0.263
7	0.952	0.487	0.000	0.487	0.002	0.487	0.004	0.239
8	1.084	0.637	0.010	0.948	0.142	1.285	0.274	0.396
9	0.919	0.977	0.166	0.898	0.086	0.818	0.006	0.389
10	1.100	0.558	0.014	0.790	0.156	1.021	0.298	0.334
11	0.813	1.444	0.626	1.022	0.336	0.600	0.046	0.393
12	0.954	0.487	0.006	0.487	0.003	0.487	0.001	0.239
13	1.030	0.577	0.085	0.620	0.046	0.573	0.007	0.266
14	1.834	0.164	0.048	0.559	0.024	0.955	0.000	0.092
15	0.808	0.756	0.046	0.521	0.028	0.286	0.010	0.206
16	1.153	0.760	0.280	1.054	0.391	1.348	0.502	0.411
17	0.798	0.792	0.081	0.533	0.045	0.273	0.009	0.204
18	1.926	0.057	0.010	0.504	0.007	0.960	0.004	0.040
19	1.400	0.443	0.061	0.624	0.084	0.805	0.107	0.235
20	1.378	0.503	0.106	0.616	0.101	0.730	0.095	0.239

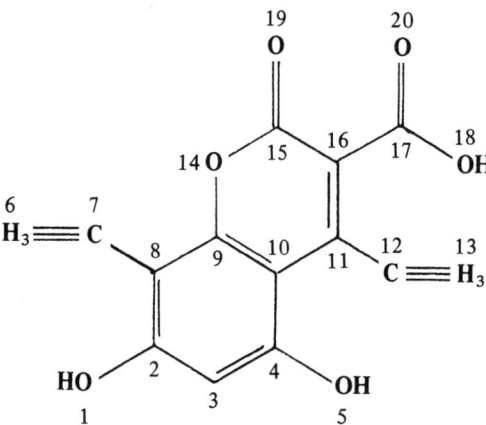

TABLE 4(S).—Aflatoxin G.

Atom	P_{rr}	SDN_r	FOD_r	SDR_r	FRD_r	SDE_r	FED_r	π_{rr}
1	1.918	0.084	0.027	0.570	0.058	1.056	0.089	0.051
2	0.924	0.941	0.220	0.872	0.221	0.804	0.222	0.375
3	1.127	0.640	0.000	0.923	0.000	1.206	0.000	0.405
4	0.913	0.984	0.209	0.915	0.210	0.846	0.212	0.389
5	1.915	0.089	0.026	0.576	0.055	1.062	0.082	0.053
6	0.920	0.355	0.001	0.378	0.012	0.400	0.022	0.173
7	1.080	0.276	0.000	0.333	0.001	0.389	0.003	0.161
8	1.084	0.637	0.010	0.948	0.142	1.258	0.274	0.396
9	0.924	0.961	0.178	0.893	0.092	0.824	0.005	0.388
10	1.101	0.556	0.019	0.789	0.160	1.022	0.301	0.333
11	0.804	1.431	0.640	1.015	0.343	0.600	0.046	0.393
12	1.080	0.276	0.000	0.331	0.000	0.385	0.000	0.161
13	0.902	0.410	0.056	0.380	0.025	0.351	0.004	0.173
14	1.835	0.160	0.050	0.559	0.025	0.957	0.000	0.091
15	0.808	0.753	0.046	0.520	0.028	0.286	0.010	0.206
16	1.161	0.724	0.270	1.040	0.389	1.357	0.507	0.404
17	0.798	0.788	0.083	0.531	0.046	0.273	0.009	0.204
18	1.926	0.056	0.010	0.504	0.007	0.952	0.004	0.040
19	1.401	0.437	0.059	0.621	0.083	0.806	0.108	0.234
20	1.379	0.494	0.106	0.613	0.101	0.731	0.096	0.238

TABLE 5(I).—3-Aminopyridoxal.

Atom	P_{rr}	SDN_r	FOD_r	SDR_r	FRD_r	SDE_r	FED_r	π_{rr}
1	1.212	0.762	0.433	0.843	0.225	0.924	0.018	0.379
2	0.842	0.791	0.116	1.035	0.276	1.280	0.437	0.414
3	0.974	0.829	0.320	0.880	0.279	0.932	0.238	0.370
4	1.129	0.625	0.269	0.853	0.168	1.080	0.067	0.357
5	0.818	0.829	0.106	0.910	0.142	0.991	0.177	0.396
6	1.090	0.723	0.185	1.053	0.307	1.384	0.430	0.426
7	0.794	0.871	0.241	0.575	0.122	0.278	0.002	0.217
8	1.303	0.590	0.248	0.601	0.132	0.612	0.015	0.248
9	1.840	0.156	0.082	1.198	0.349	2.240	0.615	0.140

TABLE 6(I).—3-Aminopyridoxal, protonated 1.

Atom	P_{π}	SDN_r	FOD_r	SDR_r	FRD_r	SDE_r	FED_r	π_{π}
1	1.647	0.617	0.240	0.656	0.132	0.696	0.014	0.183
2	0.575	1.566	0.409	1.134	0.350	0.701	0.293	0.390
3	0.932	1.092	0.282	0.924	0.236	0.756	0.190	0.371
4	1.050	1.028	0.339	0.997	0.262	0.966	0.185	0.400
5	0.769	0.857	0.004	0.800	0.041	0.744	0.078	0.366
6	1.145	1.133	0.342	1.258	0.428	1.382	0.513	0.526
7	0.792	0.924	0.108	0.602	0.056	0.279	0.005	0.218
8	1.287	0.714	0.158	0.652	0.100	0.590	0.042	0.252
9	1.804	0.297	0.108	1.116	0.395	2.024	0.681	0.195

TABLE 7(I).—3-Aminopyridoxal, protonated 2.

Atom	P_{rr}	SDN_r	FOD_r	SDR_r	FRD_r	SDE_r	FED_r	π_{rr}
1	1.175	0.723	0.033	0.790	0.236	0.857	0.439	0.421
2	0.295	0.144	0.543	0.015	0.294	0.174	0.044	0.364
3	0.581	0.203	0.383	0.253	0.253	0.304	0.124	0.383
4	1.106	0.650	0.004	0.798	0.409	0.946	0.814	0.360
5	0.468	0.232	0.350	0.299	0.294	0.366	0.239	0.467
6	0.576	−0.092	0.495	0.124	0.263	0.337	0.031	0.439
7	0.790	0.881	0.000	0.577	0.001	0.273	0.002	0.216
8	1.295	0.603	0.001	0.592	0.094	0.581	0.187	0.247
9	1.714	−0.237	0.191	0.276	0.156	0.788	0.122	0.275

TABLE 8(I).—3-Aminopyridoxal, double protonated.

Atom	P_π	SDN_r	FOD_r	SDR_r	FRD_r	SDE_r	FED_r	π_π
1	1.643	0.540	0.008	0.601	0.067	0.661	0.126	0.180
2	0.194	0.645	0.432	0.357	0.216	0.069	0.000	0.103
3	0.562	0.304	0.401	0.290	0.268	0.277	0.135	0.344
4	0.984	1.016	0.030	0.871	0.407	0.726	0.785	0.405
5	0.521	0.292	0.308	0.353	0.391	0.413	0.474	0.467
6	0.342	0.236	0.610	0.199	0.344	0.162	0.078	0.266
7	0.787	0.937	0.001	0.603	0.000	0.269	0.000	0.216
8	1.269	0.723	0.007	0.628	0.101	0.534	0.196	0.253
9	1.714	−0.237	0.191	0.276	0.156	0.788	0.120	0.275

TABLE 9.—Angelicin.

Atom	P_{rr}	SDN_r	FOD_r	SDR_r	FRD_r	SDE_r	FED_r	π_{rr}
1	1.399	0.475	0.128	0.653	0.108	0.830	0.086	0.240
2	0.813	0.766	0.106	0.530	0.058	0.294	0.010	0.207
3	1.826	0.180	0.061	0.570	0.078	0.960	0.095	0.098
4	0.919	1.015	0.157	0.999	0.262	0.982	0.367	0.418
5	1.071	0.608	0.033	0.746	0.060	0.884	0.086	0.330
6	1.113	0.655	0.007	0.938	0.165	1.222	0.321	0.401
7	0.978	0.956	0.016	1.094	0.181	1.233	0.346	0.489
8	1.786	0.212	0.036	0.562	0.059	0.912	0.083	0.118
9	0.976	0.832	0.176	0.872	0.095	0.912	0.123	0.383
10	1.058	0.755	0.008	0.916	0.126	1.078	0.244	0.411
11	1.007	0.955	0.227	0.995	0.182	1.035	0.138	0.443
12	1.062	0.602	0.043	0.763	0.066	0.925	0.090	0.331
13	0.911	1.211	0.552	0.975	0.279	0.739	0.006	0.423
14	1.081	0.968	0.449	1.129	0.282	1.291	0.116	0.490

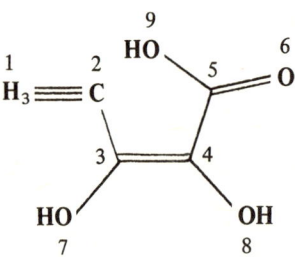

TABLE 10(P).—Ascorbic Acid.

Atom	P_{rr}	SDN_r	FOD_r	SDR_r	FRD_r	SDE_r	FED_r	π_{rr}
1	1.043	0.524	0.139	0.590	0.109	0.656	0.079	0.267
2	0.954	0.486	0.028	0.486	0.016	0.486	0.005	0.239
3	0.907	1.023	0.806	1.154	0.675	1.284	0.545	0.483
4	1.130	0.637	0.340	1.175	0.545	1.713	0.751	0.421
5	0.805	0.775	0.288	0.535	0.153	0.296	0.019	0.218
6	1.383	0.470	0.261	0.637	0.202	0.805	0.123	0.240
7	1.910	0.092	0.078	0.619	0.134	1.145	0.191	0.058
8	1.941	0.049	0.033	0.658	0.148	1.267	0.262	0.036
9	1.926	0.054	0.028	0.505	0.017	0.957	0.007	0.040

TABLE 10(S).—Ascorbic Acid.

Atom	P_{rr}	SDN_r	FOD_r	SDR_r	FRD_r	SDE_r	FED_r	π_{rr}
1	0.917	0.333	0.001	0.333	0.001	0.333	0.000	0.165
2	1.038	0.281	0.000	0.337	0.000	0.392	0.000	0.165
3	0.896	1.052	0.921	1.182	0.755	1.313	0.589	0.511
4	1.140	0.597	0.314	1.135	0.536	1.674	0.759	0.405
5	0.805	0.772	0.333	0.532	0.175	0.293	0.018	0.207
6	1.384	0.461	0.286	0.629	0.215	0.797	0.144	0.239
7	1.906	0.094	0.085	0.620	0.148	1.146	0.211	0.060
8	1.944	0.044	0.029	0.653	0.150	1.262	0.272	0.033
9	1.926	0.053	0.031	0.505	0.019	0.956	0.006	0.040

TABLE 11.—8-Azaadenine.

Atom	P_{rr}	SDN_r	FOD_r	SDR_r	FRD_r	SDE_r	FED_r	π_{rr}
1	1.802	0.204	0.099	1.043	0.297	1.882	0.496	0.169
2	0.852	1.014	0.380	0.816	0.258	0.617	0.137	0.342
3	1.282	0.584	0.002	0.890	0.099	1.196	0.196	0.371
4	0.884	1.040	0.313	0.879	0.206	0.719	0.098	0.396
5	1.369	0.648	0.227	0.954	0.298	1.260	0.369	0.388
6	0.947	0.781	0.000	0.756	0.028	0.732	0.056	0.348
7	1.563	0.430	0.145	0.835	0.146	1.239	0.148	0.278
8	1.108	0.990	0.547	0.993	0.369	0.995	0.191	0.449
9	1.211	0.734	0.268	0.892	0.214	1.049	0.161	0.395
10	1.082	0.574	0.019	0.772	0.084	0.969	0.149	0.335

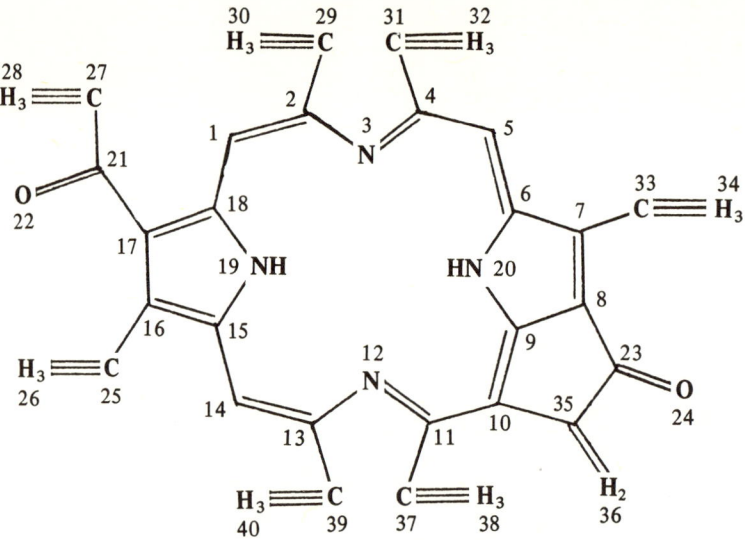

TABLE 12(P).—Bacteriopheophytin A.

Atom	P_{rr}	SDN_r	FOD_r	SDR_r	FRD_r	SDE_r	FED_r	π_{rr}
1	1.077	1.262	0.164	1.446	0.124	1.629	0.085	0.517
2	0.925	1.007	0.009	1.068	0.009	1.128	0.065	0.413
3	1.244	1.190	0.177	1.368	0.166	1.546	0.156	0.466
4	0.932	1.025	0.029	1.095	0.061	1.164	0.093	0.421
5	1.065	1.286	0.154	1.474	0.212	1.662	0.269	0.525
6	0.997	0.936	0.069	1.006	0.047	1.075	0.026	0.378
7	0.988	1.285	0.150	1.257	0.136	1.230	0.122	0.464
8	1.103	0.948	0.105	1.049	0.053	1.150	0.000	0.390
9	0.989	0.924	0.032	1.006	0.079	1.089	0.125	0.387
10	1.023	1.313	0.176	1.439	0.130	1.565	0.085	0.501
11	0.937	1.006	0.014	1.092	0.035	1.178	0.055	0.426
12	1.232	1.252	0.192	1.374	0.171	1.496	0.150	0.469
13	0.937	1.004	0.025	1.095	0.053	1.186	0.080	0.422
14	1.055	1.343	0.170	1.472	0.212	1.600	0.253	0.527
15	1.000	0.920	0.065	1.011	0.043	1.102	0.021	0.378
16	0.986	1.312	0.147	1.255	0.135	1.218	0.111	0.465
17	1.104	0.927	0.101	1.050	0.051	1.171	0.000	0.390
18	0.970	0.931	0.035	0.992	0.073	1.052	0.111	0.380
19	1.637	0.362	0.002	0.952	0.002	1.541	0.054	0.245
20	1.633	0.374	0.004	0.961	0.030	1.547	0.056	0.250
21	0.810	0.894	0.019	0.587	0.010	0.280	0.000	0.280
22	1.318	0.710	0.037	0.691	0.019	0.672	0.000	0.260
23	0.809	0.901	0.018	0.592	0.009	0.283	0.000	0.209
24	1.311	0.716	0.035	0.692	0.017	0.667	0.000	0.261
25	0.952	0.487	0.000	0.487	0.000	0.487	0.000	0.239
26	1.048	0.561	0.020	0.605	0.017	0.450	0.015	0.266
27	0.958	0.484	0.000	0.484	0.000	0.484	0.000	0.235
28	1.007	0.594	0.005	0.567	0.002	0.541	0.000	0.268
29	0.953	0.487	0.000	0.487	0.000	0.487	0.000	0.239
30	1.043	0.525	0.001	0.583	0.005	0.690	0.008	0.265
31	0.953	0.487	0.000	0.487	0.000	0.487	0.000	0.239

TABLE 12(P).—*Continued.*

32	1.044	0.537	0.003	0.586	0.008	0.644	0.012	0.265
33	0.952	0.487	0.000	0.487	0.000	0.487	0.000	0.239
34	1.048	0.588	0.018	0.605	0.017	0.651	0.016	0.266
35	0.962	0.473	0.000	0.473	0.000	0.473	0.000	0.226
36	1.008	0.637	0.006	0.639	0.009	0.641	0.011	0.283
37	0.953	0.487	0.000	0.487	0.000	0.487	0.000	0.239
38	1.044	0.525	0.003	0.585	0.004	0.646	0.007	0.265
39	0.953	0.487	0.000	0.487	0.000	0.487	0.000	0.239
40	1.044	0.525	0.003	0.586	0.007	0.647	0.011	0.265

TABLE 12(S).—Bacteriopheophytin A.

Atom	$P_{\pi\pi}$	SDN_r	FOD_r	SDR_r	FRD_r	SDE_r	FED_r	$\pi_{\pi\pi}$
1	1.086	1.266	0.165	1.431	0.106	1.657	0.047	0.510
2	0.919	1.003	0.009	1.083	0.009	1.163	0.074	0.422
3	1.251	1.128	0.175	1.345	0.147	1.562	0.119	0.457
4	0.926	1.019	0.030	1.112	0.077	1.205	0.124	0.425
5	1.074	1.235	0.157	1.470	0.201	1.705	0.244	0.521
6	1.002	0.921	0.073	1.014	0.061	1.107	0.049	0.378
7	0.982	1.269	0.154	1.270	0.148	1.271	0.141	0.470
8	1.109	0.900	0.103	1.030	0.053	1.160	0.004	0.384
9	0.988	0.895	0.031	1.008	0.088	1.122	0.146	0.385
10	1.023	1.273	0.180	1.435	0.113	1.596	0.047	0.501
11	0.934	0.998	0.014	1.110	0.049	1.222	0.084	0.430
12	1.236	1.199	0.193	1.352	0.193	1.505	0.115	0.462
13	0.932	0.996	0.026	1.113	0.069	1.231	0.112	0.426
14	1.062	1.299	0.175	1.415	0.203	1.630	0.230	0.523
15	1.006	0.905	0.069	1.023	0.056	1.140	0.043	0.379
16	0.980	1.289	0.161	1.273	0.146	1.256	0.131	0.471
17	1.113	0.882	0.098	1.035	0.051	1.188	0.003	0.385
18	0.971	0.922	0.037	1.002	0.085	1.082	0.132	0.382
19	1.639	0.351	0.003	0.947	0.003	1.548	0.037	0.243
20	1.636	0.364	0.005	0.961	0.022	1.559	0.039	0.247
21	0.801	0.878	0.017	0.580	0.009	0.281	0.000	0.209
22	1.320	0.555	0.034	0.660	0.017	0.664	0.001	0.253
23	0.802	0.887	0.019	0.586	0.010	0.284	0.000	0.210
24	1.315	0.675	0.037	0.671	0.019	0.666	0.001	0.255
25	1.079	0.277	0.000	0.332	0.000	0.388	0.000	0.161
26	0.913	0.401	0.011	0.399	0.011	0.398	0.010	0.173
27	1.081	0.274	0.000	0.329	0.000	0.383	0.000	0.159
28	0.894	0.399	0.002	0.366	0.001	0.333	0.000	0.172
29	1.080	0.276	0.000	0.332	0.000	0.387	0.000	0.161
30	0.910	0.380	0.001	0.386	0.004	0.692	0.007	0.173
31	1.080	0.276	0.000	0.332	0.000	0.388	0.000	0.161
32	0.911	0.382	0.002	0.388	0.006	0.395	0.009	0.173
33	1.079	0.277	0.000	0.332	0.000	0.388	0.001	0.161
34	0.913	0.399	0.011	0.399	0.011	0.399	0.011	0.173
35	1.078	0.269	0.000	0.324	0.000	0.379	0.000	0.155
36	0.889	0.434	0.004	0.417	0.004	0.400	0.004	0.180
37	1.080	0.276	0.000	0.322	0.000	0.388	0.001	0.161
38	0.911	0.380	0.002	0.388	0.004	0.396	0.006	0.173
39	1.080	0.276	0.000	0.322	0.000	0.388	0.000	0.161
40	0.911	0.380	0.002	0.388	0.005	0.397	0.008	0.173

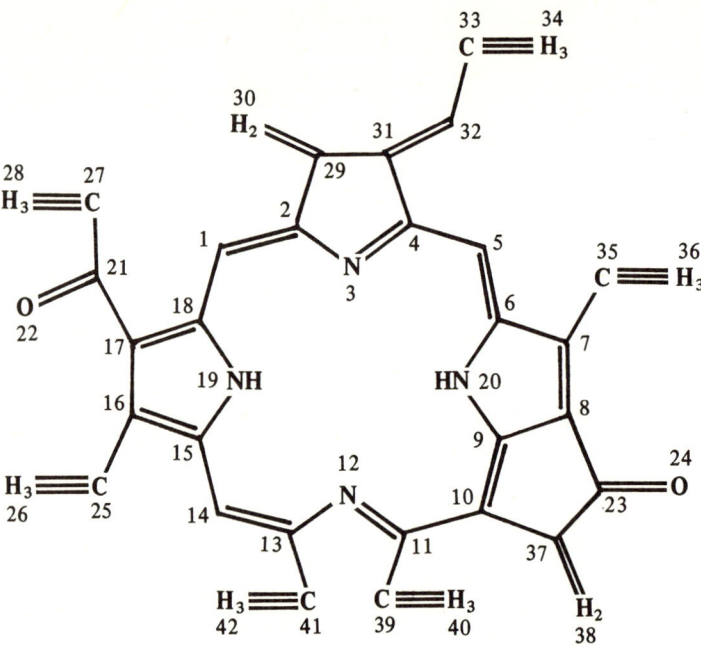

TABLE 13(P).—Bacteriopheophytin *B*.

Atom	P_{rr}	SDN_r	FOD_r	SDR_r	FRD_r	SDE_r	FED_r	π_{rr}
1	1.076	1.306	0.172	1.489	0.086	1.673	0.000	0.527
2	0.928	0.997	0.008	1.058	0.065	1.118	0.122	0.413
3	1.258	1.190	0.183	1.406	0.099	1.621	0.015	0.480
4	0.984	0.945	0.032	1.015	0.085	1.084	0.139	0.371
5	1.044	1.377	0.130	1.565	0.152	1.754	0.173	0.565
6	1.000	0.920	0.068	0.990	0.070	1.059	0.082	0.373
7	0.980	1.303	0.133	1.275	0.135	1.248	0.137	0.467
8	1.102	0.944	0.101	1.045	0.058	1.146	0.016	0.390
9	0.979	0.936	0.027	1.019	0.083	1.101	0.129	0.388
10	1.024	1.315	0.175	1.441	0.091	1.568	0.006	0.501
11	0.935	1.015	0.011	1.101	0.064	1.188	0.117	0.427
12	1.231	1.256	0.188	1.378	0.116	1.500	0.044	0.470
13	0.935	1.022	0.028	1.113	0.084	1.204	0.139	0.423
14	1.055	1.348	0.166	1.477	0.145	1.606	0.124	0.527
15	0.998	0.938	0.068	1.029	0.075	1.120	0.081	0.380
16	0.985	1.325	0.159	1.278	0.141	1.231	0.122	0.466
17	1.103	0.947	0.104	1.068	0.064	1.189	0.023	0.392
18	0.969	0.933	0.034	0.993	0.080	1.054	0.125	0.380
19	1.637	0.366	0.003	0.956	0.005	1.546	0.007	0.246
20	1.634	0.387	0.005	0.974	0.010	1.560	0.016	0.251
21	0.810	0.896	0.019	0.589	0.010	0.282	0.001	0.208
22	1.318	0.715	0.038	0.696	0.022	0.677	0.005	0.260
23	0.809	0.900	0.017	0.591	0.009	0.283	0.001	0.209
24	1.311	0.716	0.033	0.691	0.018	0.666	0.004	0.261
25	0.952	0.487	0.000	0.487	0.000	0.487	0.000	0.239
26	1.047	0.563	0.020	0.607	0.018	0.651	0.016	0.266
27	0.958	0.399	0.001	0.366	0.001	0.333	0.000	0.172

TABLE 13(P).—*Continued*.

28	1.009	0.595	0.005	0.568	0.003	0.541	0.000	0.268
29	0.958	0.476	0.000	0.476	0.000	0.476	0.001	0.231
30	1.034	0.585	0.002	0.661	0.019	0.737	0.035	0.280
31	1.006	0.828	0.001	0.930	0.015	1.031	0.029	0.401
32	0.970	1.337	0.034	1.509	0.118	1.681	0.202	0.613
33	0.953	0.484	0.000	0.484	0.000	0.484	0.000	0.238
34	1.046	0.600	0.004	0.629	0.015	0.700	0.026	0.270
35	0.952	0.487	0.000	0.487	0.000	0.487	0.000	0.239
36	1.047	0.560	0.016	0.607	0.017	0.653	0.018	0.266
37	0.962	0.473	0.000	0.473	0.000	0.473	0.000	0.226
38	1.008	0.638	0.007	0.640	0.000	0.642	0.002	0.283
39	0.953	0.487	0.000	0.487	0.000	0.487	0.000	0.239
40	1.044	0.526	0.001	0.568	0.008	0.647	0.015	0.265
41	0.953	0.487	0.000	0.487	0.000	0.487	0.000	0.239
42	1.044	0.527	0.003	0.588	0.011	0.649	0.018	0.265

TABLE 13(S).—Bacteriopheophytin B.

Atom	P_{π}	SDN_r	FOD_r	SDR_r	FRD_r	SDE_r	FED_r	π_{π}
1	1.086	1.253	0.174	1.482	0.088	1.711	0.001	0.522
2	0.920	0.989	0.008	1.070	0.070	1.150	0.133	0.414
3	1.254	1.251	0.185	1.429	0.104	1.607	0.023	0.487
4	0.984	0.932	0.033	1.025	0.089	1.118	0.146	0.370
5	1.050	1.333	0.131	1.572	0.145	1.810	0.159	0.564
6	1.005	0.904	0.072	0.998	0.077	1.091	0.082	0.372
7	0.974	1.288	0.136	1.271	0.138	1.249	0.140	0.474
8	1.109	0.896	0.098	1.026	0.059	1.159	0.020	0.383
9	0.983	0.908	0.025	1.024	0.083	1.139	0.141	1.386
10	1.024	1.275	0.178	1.437	0.090	1.598	0.002	0.501
11	0.932	1.008	0.011	1.122	0.069	1.235	0.127	0.431
12	1.237	1.203	0.189	1.356	0.111	1.509	0.033	0.463
13	0.930	1.015	0.030	1.134	0.089	1.253	0.149	0.437
14	1.062	1.306	0.171	1.471	0.141	1.637	0.111	0.524
15	1.004	0.925	0.073	1.045	0.084	1.164	0.094	0.381
16	0.979	1.304	0.161	1.289	0.143	1.273	0.125	0.472
17	1.112	0.901	0.102	1.056	0.065	1.211	0.028	0.387
18	0.970	0.924	0.036	0.994	0.081	1.085	0.126	0.380
19	1.640	0.355	0.003	0.954	0.003	1.553	0.009	0.243
20	1.637	0.377	0.006	0.975	0.009	1.573	0.011	0.248
21	0.801	0.880	0.018	0.562	0.010	0.283	0.002	0.209
22	1.320	0.661	0.035	0.665	0.021	0.670	0.007	0.253
23	0.802	0.887	0.019	0.586	0.010	0.284	0.001	0.210
24	1.315	0.674	0.036	0.670	0.020	0.665	0.004	0.255
25	1.079	0.277	0.000	0.332	0.000	0.388	0.001	0.161
26	0.913	0.402	0.011	0.401	0.010	0.399	0.009	0.173
27	1.081	0.274	0.000	0.329	0.000	0.383	0.000	0.159
28	0.894	0.484	0.000	0.484	0.000	0.484	0.000	0.235
29	1.008	0.271	0.000	0.327	0.001	0.383	0.001	0.156
30	0.901	0.418	0.001	0.439	0.012	0.460	0.022	0.181
31	1.012	0.788	0.000	0.929	0.018	1.069	0.035	0.400
32	0.976	1.193	0.033	1.548	0.129	1.803	0.225	0.621
33	1.079	0.276	0.000	0.332	0.001	0.388	0.001	0.160

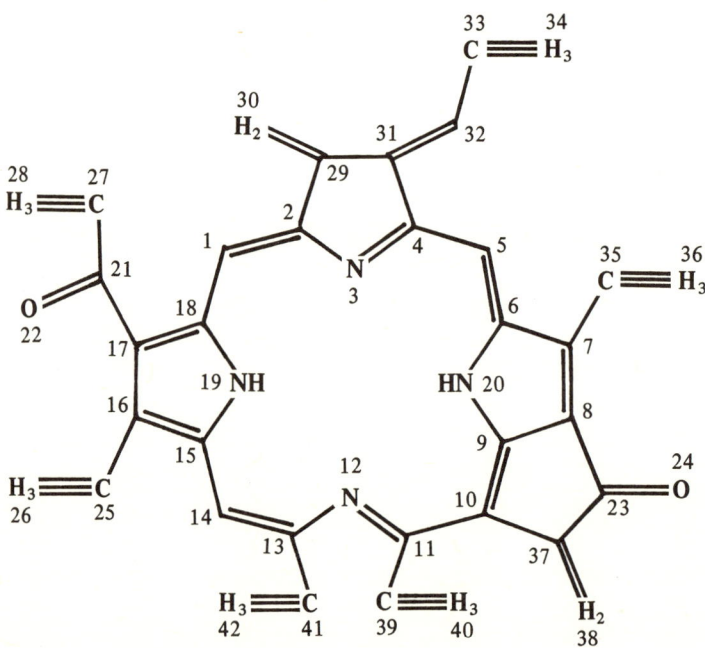

TABLE 13(S).—*Continued.*

34	0.913	0.313	0.002	0.417	0.009	0.435	0.017	0.175
35	1.079	0.277	0.000	0.332	0.000	0.388	0.001	0.161
36	0.913	0.400	0.010	0.401	0.010	0.401	0.010	0.173
37	1.078	0.269	0.000	0.324	0.000	0.379	0.000	0.155
38	0.890	0.434	0.004	0.417	0.002	0.401	0.001	0.180
39	1.080	0.276	0.000	0.332	0.000	0.388	0.001	0.161
40	0.911	0.381	0.001	0.389	0.005	0.397	0.009	0.173
41	1.080	0.276	0.000	0.332	0.000	0.388	0.000	0.161
42	0.911	0.381	0.002	0.390	0.007	0.398	0.011	0.173

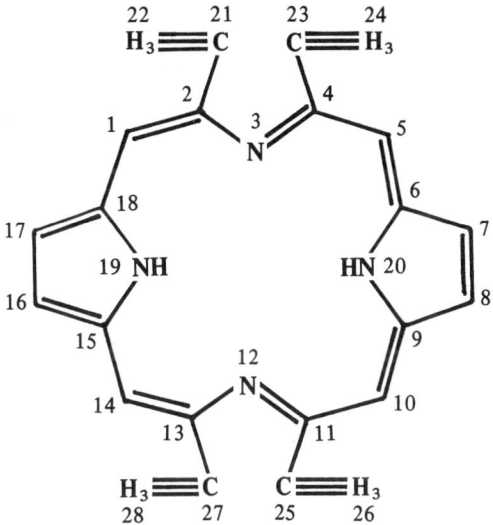

TABLE 14(P).—Bacteriochlorin.

Atom	P_{π}	SDN_r	FOD_r	SDR_r	FRD_r	SDE_r	FED_r	π_{π}
1	1.064	1.251	0.179	1.425	0.107	1.598	0.036	0.521
2	0.935	0.988	0.022	1.081	0.108	1.173	0.196	0.419
3	1.239	1.168	0.205	1.133	0.102	1.499	0.000	0.464
4	0.935	0.988	0.022	1.081	0.108	1.173	0.196	0.419
5	1.064	1.251	0.179	1.425	0.107	1.598	0.036	0.521
6	0.994	0.903	0.068	0.996	0.110	1.088	0.151	0.375
7	1.067	1.053	0.125	1.145	0.108	1.238	0.091	0.450
8	1.067	1.053	0.125	1.145	0.108	1.238	0.091	0.450
9	0.994	0.903	0.068	0.996	0.110	1.088	0.151	0.375
10	1.064	1.251	0.179	1.425	0.107	1.598	0.036	0.521
11	0.935	0.988	0.022	1.081	0.108	1.173	0.196	0.419
12	1.239	1.168	0.205	1.133	0.102	1.499	0.000	0.464
13	0.935	0.988	0.022	1.081	0.108	1.173	0.196	0.419
14	1.064	1.251	0.179	1.425	0.107	1.598	0.036	0.521
15	0.994	0.903	0.068	0.996	0.110	1.088	0.151	0.375
16	1.067	1.053	0.125	1.145	0.108	1.238	0.091	0.450
17	1.067	1.053	0.125	1.145	0.108	1.238	0.091	0.450
18	0.994	0.903	0.068	0.996	0.110	1.088	0.151	0.375
19	1.645	0.332	0.000	0.929	0.000	1.526	0.000	0.236
20	1.645	0.332	0.000	0.929	0.000	1.526	0.000	0.236
21	0.953	0.487	0.000	0.487	0.000	0.487	0.001	0.239
22	1.044	0.523	0.003	0.584	0.014	0.646	0.030	0.265
23	0.953	0.487	0.000	0.487	0.000	0.487	0.001	0.239
24	1.044	0.523	0.003	0.584	0.014	0.646	0.030	0.265
25	0.953	0.487	0.000	0.487	0.000	0.487	0.001	0.239
26	1.044	0.523	0.003	0.584	0.014	0.646	0.030	0.265
27	0.953	0.487	0.000	0.487	0.000	0.487	0.001	0.239
28	1.044	0.523	0.003	0.584	0.014	0.646	0.030	0.265

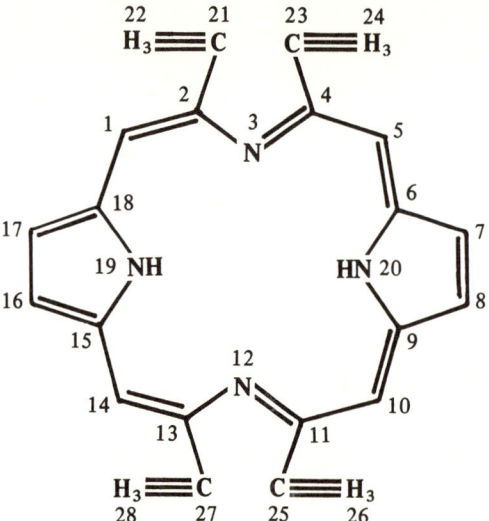

TABLE 14(S).—Bacteriochlorin.

Atom	P_π	SDN_r	FOD_r	SDR_r	FRD_r	SDE_r	FED_r	π_π
1	1.072	1.227	0.181	1.421	0.110	1.616	0.038	0.517
2	0.928	0.989	0.022	1.095	0.111	1.201	0.199	0.423
3	1.245	1.134	0.200	1.317	0.100	1.500	0.000	0.457
4	0.928	0.989	0.022	1.095	0.111	1.201	0.199	0.423
5	1.072	1.227	0.181	1.421	0.110	1.616	0.038	0.517
6	0.995	0.898	0.068	1.004	0.110	1.110	0.153	0.376
7	1.070	1.045	0.127	1.150	0.110	1.256	0.093	0.451
8	1.070	1.045	0.127	1.150	0.110	1.256	0.093	0.451
9	0.995	0.898	0.068	1.004	0.110	1.110	0.153	0.376
10	1.072	1.227	0.181	1.421	0.110	1.616	0.038	0.517
11	0.928	0.989	0.022	1.095	0.111	1.201	0.199	0.423
12	1.245	1.134	0.200	1.317	0.100	1.500	0.000	0.457
13	0.928	0.989	0.022	1.095	0.111	1.201	0.199	0.423
14	1.072	1.227	0.181	1.421	0.110	1.616	0.038	0.517
15	0.995	0.898	0.068	1.004	0.110	1.110	0.153	0.376
16	1.070	1.045	0.127	1.150	0.110	1.256	0.093	0.451
17	1.070	1.045	0.127	1.150	0.110	1.256	0.093	0.451
18	0.995	0.898	0.068	1.004	0.110	1.110	0.153	0.376
19	1.646	0.324	0.000	0.926	0.000	1.527	0.000	0.234
20	1.646	0.324	0.000	0.926	0.000	1.527	0.000	0.234
21	1.080	0.276	0.000	0.332	0.001	0.388	0.001	0.161
22	0.911	0.380	0.002	0.387	0.008	0.395	0.015	0.173
23	1.080	0.276	0.000	0.332	0.001	0.388	0.001	0.161
24	0.911	0.380	0.002	0.387	0.008	0.395	0.015	0.173
25	1.080	0.276	0.000	0.332	0.001	0.388	0.001	0.161
26	0.911	0.380	0.002	0.387	0.008	0.395	0.015	0.173
27	1.080	0.276	0.000	0.332	0.001	0.388	0.001	0.161
28	0.911	0.380	0.002	0.387	0.008	0.395	0.015	0.173

TABLE 15.—Benzanthracene.

Atom	P_{rr}	SDN_r	FOD_r	SDR_r	FRD_r	SDE_r	FED_r	π_{rr}
1	1.000	0.902	0.075	0.902	0.075	0.902	0.075	0.409
2	1.000	1.054	0.209	1.054	0.209	1.054	0.209	0.452
3	1.000	0.713	0.005	0.713	0.005	0.713	0.005	0.331
4	1.000	1.251	0.396	1.251	0.396	1.251	0.396	0.513
5	1.000	0.751	0.047	0.751	0.047	0.751	0.047	0.336
6	1.000	1.039	0.177	1.039	0.177	1.039	0.177	0.448
7	1.000	1.040	0.166	1.040	0.166	1.040	0.166	0.449
8	1.000	0.749	0.056	0.749	0.056	0.749	0.056	0.336
9	1.000	0.980	0.051	0.980	0.051	0.980	0.051	0.439
10	1.000	0.868	0.018	0.868	0.018	0.868	0.018	0.404
11	1.000	0.908	0.082	0.908	0.082	0.908	0.082	0.410
12	1.000	0.935	0.000	0.935	0.000	0.935	0.000	0.429
13	1.000	0.811	0.083	0.811	0.083	0.811	0.083	0.353
14	1.000	0.791	0.012	0.791	0.012	0.791	0.012	0.351
15	1.000	1.187	0.309	1.187	0.309	1.187	0.309	0.495
16	1.000	0.729	0.020	0.729	0.020	0.729	0.020	0.333
17	1.000	1.040	0.181	1.040	0.181	1.040	0.181	0.449
18	1.000	0.915	0.111	0.915	0.111	0.915	0.111	0.410

TABLE 16(P).—Bilirubin.

Atom	P_π	SDN_r	FOD_r	SDR_r	FRD_r	SDE_r	FED_r	π_π
1	1.079	1.310	0.117	1.742	0.188	2.175	0.259	0.679
2	0.988	0.875	0.015	0.874	0.014	0.874	0.013	0.399
3	1.114	0.899	0.089	1.331	0.162	1.763	0.234	0.426
4	0.945	1.355	0.147	1.182	0.074	1.008	0.001	0.466
5	1.020	0.794	0.033	1.161	0.129	1.528	0.224	0.387
6	0.993	1.526	0.188	1.526	0.187	1.526	0.186	0.557
7	1.057	0.617	0.001	0.984	0.072	1.351	0.143	0.358
8	1.065	0.927	0.075	1.256	0.118	1.586	0.162	0.448
9	1.091	0.606	0.003	0.936	0.030	1.265	0.056	0.376
10	1.029	1.006	0.066	1.373	0.151	1.740	0.236	0.482
11	1.001	0.475	0.001	0.475	0.001	0.475	0.001	0.230
12	1.040	0.955	0.068	1.366	0.041	1.777	0.015	0.478
13	1.092	0.601	0.077	0.938	0.004	1.275	0.004	0.375
14	1.076	0.869	0.126	1.239	0.044	1.609	0.010	0.441
15	1.058	0.622	0.001	1.000	0.005	1.378	0.010	0.360
16	1.018	1.369	0.194	1.451	0.102	1.532	0.010	0.537
17	1.009	0.835	0.034	1.213	0.025	1.591	0.015	0.406
18	1.006	1.038	0.126	0.933	0.063	0.828	0.000	0.367
19	1.075	1.304	0.209	1.682	0.112	2.060	0.014	0.598
20	0.851	0.798	0.047	0.624	0.025	0.450	0.002	0.222
21	1.416	0.628	0.048	0.978	0.044	1.329	0.009	0.273
22	0.840	0.783	0.024	0.609	0.030	0.435	0.036	0.219
23	1.408	0.562	0.040	0.912	0.095	1.263	0.150	0.260
24	1.752	0.151	0.000	1.041	0.054	1.932	0.108	0.151
25	1.589	0.478	0.036	0.885	0.026	1.292	0.016	0.270
26	1.011	0.632	0.035	0.727	0.041	0.822	0.046	0.287
27	1.591	0.460	0.037	0.875	0.019	1.291	0.001	0.267
28	1.742	0.172	0.000	1.056	0.004	1.940	0.007	0.162
29	0.952	0.487	0.000	0.487	0.000	0.487	0.000	0.239
30	1.044	0.566	0.019	0.595	0.009	0.623	0.000	0.267
31	0.952	0.487	0.000	0.487	0.000	0.487	0.000	0.239
32	1.056	0.515	0.009	0.605	0.015	0.696	0.021	0.265
33	0.952	0.488	0.000	0.488	0.000	0.488	0.000	0.239
34	1.059	0.478	0.000	0.569	0.004	0.660	0.000	0.262
35	0.952	0.488	0.000	0.488	0.000	0.488	0.000	0.239

TABLE 16(P).—*Continued.*

36	1.059	0.478	0.000	0.569	0.004	0.660	0.001	0.262
37	0.952	0.487	0.000	0.487	0.000	0.487	0.000	0.239
38	1.059	0.508	0.010	0.603	0.005	0.699	0.001	0.265
39	0.996	0.898	0.021	0.898	0.011	0.898	0.000	0.409
40	0.993	1.510	0.166	1.005	0.083	1.300	0.000	0.628
41	0.952	0.485	0.001	0.485	0.000	0.485	0.000	0.238
42	1.057	0.557	0.026	0.653	0.014	0.749	0.002	0.270

TABLE 16(S).—Bilirubin.

Atom	P_π	SDN_r	FOD_r	SDR_r	FRD_r	SDE_r	FED_r	π_{rr}
1	1.086	1.260	0.095	1.734	0.183	2.209	0.271	0.672
2	0.988	0.875	0.013	0.875	0.013	0.875	0.013	0.399
3	1.123	0.846	0.071	1.321	0.158	1.795	0.245	0.418
4	0.945	1.349	0.128	1.181	0.065	1.012	0.001	0.471
5	1.026	0.778	0.029	1.167	0.131	1.555	0.234	0.386
6	0.989	1.510	0.165	1.517	0.178	1.524	0.192	0.556
7	1.065	0.601	0.000	0.990	0.075	1.378	0.149	0.356
8	1.063	0.923	0.066	1.264	0.117	1.605	0.168	0.452
9	1.091	0.594	0.003	0.932	0.029	1.270	0.055	0.376
10	1.036	0.999	0.060	1.390	0.153	1.782	0.247	0.488
11	1.000	0.321	0.000	0.321	0.000	0.321	0.000	0.558
12	1.049	0.939	0.081	1.393	0.045	1.848	0.009	0.483
13	1.093	0.589	0.090	0.936	0.003	1.284	0.002	0.375
14	1.076	0.867	0.090	1.254	0.048	1.652	0.006	0.444
15	1.066	0.606	0.000	1.011	0.003	1.414	0.006	0.358
16	1.021	1.331	0.222	1.451	0.114	1.571	0.006	0.536
17	1.013	0.824	0.039	1.230	0.024	1.636	0.009	0.406
18	1.013	0.989	0.139	0.913	0.069	0.837	0.000	0.364
19	1.066	1.309	0.248	1.715	0.129	2.120	0.009	0.612
20	0.852	0.796	0.058	0.628	0.030	0.460	0.001	0.222
21	1.417	0.625	0.043	0.992	0.050	1.358	0.006	0.274
22	0.840	0.777	0.021	0.608	0.029	0.440	0.038	0.219
23	1.410	0.548	0.033	0.913	0.095	1.277	0.157	0.260
24	1.754	0.148	0.000	1.047	0.056	1.946	0.112	0.149
25	1.592	0.463	0.030	0.880	0.023	1.297	0.016	0.267
26	1.007	0.422	0.020	0.482	0.023	0.543	0.025	0.186
27	1.594	0.412	0.040	0.871	0.020	1.300	0.000	0.264
28	1.743	0.171	0.001	1.062	0.003	1.953	0.004	0.161
29	1.079	0.276	0.000	0.332	0.000	0.387	0.000	0.161
30	0.911	0.404	0.009	0.392	0.005	0.389	0.000	0.173
31	1.079	0.277	0.000	0.333	0.000	0.389	0.001	0.161
32	0.919	0.375	0.005	0.399	0.009	0.424	0.012	0.173
33	1.080	0.276	0.000	0.333	0.000	0.389	0.001	0.161
34	0.921	0.353	0.000	0.377	0.002	0.401	0.004	0.172
35	1.080	0.276	0.000	0.333	0.000	0.389	0.001	0.161
36	0.921	0.353	0.000	0.377	0.002	0.401	0.004	0.172
37	1.079	0.277	0.000	0.333	0.000	0.389	0.000	0.161
38	0.920	0.371	0.006	0.399	0.003	0.427	0.000	0.173
39	0.996	0.900	0.026	0.900	0.013	0.900	0.000	0.409
40	0.998	1.464	0.187	1.388	0.094	1.313	0.000	0.625
41	1.079	0.277	0.000	0.333	0.000	0.389	0.000	0.161
42	0.918	0.401	0.017	0.430	0.009	0.459	0.001	0.175

TABLE 17(P).—Biliverdin.

Atom	P_π	SDN_r	FOD_r	SDR_r	FRD_r	SDE_r	FED_r	π_π
1	1.377	1.113	0.096	1.083	0.084	1.054	0.072	0.291
2	0.838	0.996	0.043	0.685	0.030	0.375	0.017	0.217
3	1.012	2.188	0.189	1.917	0.148	1.646	0.107	0.650
4	1.004	1.207	0.055	1.018	0.028	0.830	0.000	0.370
5	0.930	1.640	0.135	1.358	0.123	1.098	0.112	0.423
6	1.042	1.291	0.030	1.369	0.053	1.446	0.075	0.519
7	0.953	1.467	0.117	1.196	0.096	0.925	0.074	0.386
8	1.053	0.853	0.000	1.092	0.037	1.332	0.074	0.427
9	0.988	1.554	0.118	1.299	0.074	1.044	0.031	0.449
10	1.021	0.751	0.000	0.969	0.056	1.188	0.112	0.368
11	0.871	3.190	0.344	2.052	0.173	0.914	0.001	0.549
12	1.049	0.720	0.007	0.940	0.062	1.159	0.117	0.348
13	1.050	1.184	0.068	1.241	0.058	1.298	0.047	0.452
14	1.073	0.888	0.010	1.144	0.040	1.408	0.070	0.437
15	1.025	1.030	0.065	1.044	0.084	1.058	0.097	0.359
16	1.002	1.516	0.032	1.516	0.084	1.516	0.136	0.562
17	0.983	1.215	0.078	1.229	0.117	1.243	0.155	0.400
18	0.951	1.505	0.054	1.260	0.027	1.016	0.001	0.470
19	1.084	1.299	0.086	1.409	0.126	1.520	0.165	0.445
20	0.833	0.879	0.022	0.634	0.023	0.390	0.025	0.217
21	1.389	0.796	0.048	0.950	0.077	1.105	0.105	0.271
22	1.724	0.293	0.016	1.023	0.036	1.753	0.056	0.186
23	1.614	0.778	0.073	1.085	0.039	1.391	0.005	0.279
24	1.347	1.167	0.097	1.431	0.065	1.695	0.031	0.451
25	1.743	0.224	0.011	1.028	0.044	1.831	0.076	0.164
26	0.952	0.485	0.000	0.485	0.000	0.485	0.000	0.238
27	1.049	0.684	0.023	0.681	0.018	0.698	0.014	0.271
28	0.990	1.676	0.057	1.487	0.029	1.299	0.000	0.632
29	0.996	0.897	0.001	0.897	0.001	0.897	0.000	0.409
30	0.952	0.487	0.000	0.487	0.000	0.487	0.000	0.239
31	1.055	0.507	0.000	0.586	0.004	0.665	0.009	0.264
32	0.951	0.487	0.000	0.487	0.000	0.487	0.000	0.239
33	1.047	0.591	0.014	0.610	0.009	0.629	0.004	0.266
34	0.952	0.487	0.000	0.487	0.000	0.487	0.000	0.239
35	1.054	0.546	0.008	0.603	0.007	0.660	0.006	0.265
36	0.952	0.487	0.000	0.487	0.000	0.487	0.000	0.239
37	1.057	0.509	0.001	0.592	0.005	0.674	0.009	0.265

TABLE 17(P).—*Continued.*

38	0.952	0.486	0.000	0.486	0.000	0.486	0.000	0.239
39	1.044	0.584	0.007	0.604	0.003	0.624	0.000	0.267
40	0.989	0.842	0.002	0.874	0.006	0.874	0.009	0.399
41	1.049	1.709	0.090	1.819	0.137	1.930	0.183	0.699

TABLE 17(S).—Biliverdin.

Atom	P_{π}	SDN_r	FOD_r	SDR_r	FRD_r	SDE_r	FED_r	π_{π}
1	1.376	1.092	0.099	1.079	0.087	1.066	0.074	0.293
2	0.839	0.989	0.045	0.684	0.032	0.479	0.016	0.218
3	1.000	2.161	0.196	1.918	0.154	1.675	0.112	0.650
4	1.013	1.129	0.052	0.985	0.025	0.842	0.000	0.367
5	0.931	1.601	0.139	1.358	0.127	1.115	0.114	0.424
6	1.046	1.247	0.029	1.358	0.054	1.470	0.079	0.517
7	0.956	1.426	0.121	1.183	0.098	0.940	0.074	0.387
8	1.054	0.833	0.000	1.097	0.038	1.361	0.076	0.428
9	0.987	1.504	0.121	1.280	0.075	1.056	0.030	0.451
10	1.026	0.742	0.000	0.977	0.058	1.213	0.115	0.367
11	0.872	3.036	0.346	1.977	0.173	0.918	0.001	0.547
12	1.053	0.711	0.007	0.946	0.063	1.182	0.120	0.348
13	1.052	1.136	0.067	1.231	0.057	1.326	0.047	0.454
14	1.075	0.849	0.008	1.144	0.039	1.439	0.070	0.438
15	1.029	1.002	0.066	1.041	0.083	1.080	0.100	0.359
16	1.000	1.500	0.035	1.506	0.083	1.513	0.132	0.560
17	0.988	1.187	0.082	1.226	0.119	1.265	0.157	0.400
18	0.942	1.496	0.057	1.258	0.029	1.021	0.001	0.475
19	1.093	1.218	0.085	1.383	0.125	1.549	0.166	0.438
20	0.834	0.868	0.022	0.631	0.024	0.393	0.025	0.217
21	1.391	0.796	0.049	0.943	0.077	1.117	0.106	0.270
22	1.725	0.284	0.016	1.021	0.036	1.758	0.056	0.186
23	1.618	0.726	0.071	1.063	0.038	1.400	0.006	0.275
24	1.352	1.116	0.100	1.414	0.064	1.712	0.029	0.448
25	1.745	0.220	0.011	1.031	0.044	1.842	0.076	0.163
26	1.079	0.278	0.000	0.333	0.000	0.388	0.000	0.161
27	0.914	0.461	0.014	0.444	0.011	0.427	0.008	0.175
28	0.996	1.601	0.053	1.458	0.027	1.314	0.000	0.629
29	0.995	0.898	0.001	0.898	0.001	0.898	0.000	0.409
30	1.079	0.279	0.000	0.333	0.000	0.389	0.000	0.161
31	0.918	0.369	0.000	0.388	0.003	0.407	0.006	0.173
32	1.079	0.277	0.000	0.332	0.000	0.387	0.000	0.161
33	0.913	0.416	0.008	0.400	0.005	0.384	0.002	0.173
34	1.079	0.277	0.000	0.332	0.000	0.387	0.000	0.161
35	0.918	0.390	0.005	0.397	0.004	0.404	0.003	0.173
36	1.079	0.276	0.000	0.333	0.000	0.389	0.000	0.161
37	0.920	0.370	0.001	0.391	0.003	0.412	0.005	0.173
38	1.079	0.277	0.000	0.332	0.000	0.387	0.000	0.161
39	0.911	0.415	0.004	0.398	0.002	0.381	0.000	0.173
40	0.988	0.875	0.002	0.875	0.006	0.875	0.009	0.399
41	1.056	1.630	0.089	1.796	0.136	1.961	0.183	0.693

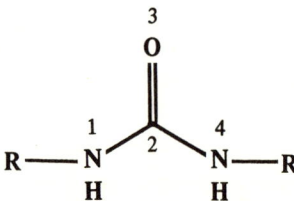

TABLE 18.—Biotin (Urea).

Atom	P_{rr}	SDN_r	FOD_r	SDR_r	FRD_r	SDE_r	FED_r	π_{rr}
1	1.872	0.073	0.128	0.903	0.564	1.733	1.000	0.084
2	0.806	0.683	1.194	0.473	0.597	0.263	0.000	0.199
3	1.450	0.315	0.550	0.564	0.275	0.813	0.000	0.210
4	1.872	0.073	0.128	0.903	0.564	1.733	1.000	0.084

TABLE 19(P).—Biopterin.

Atom	P_{rr}	SDN_r	FOD_r	SDR_r	FRD_r	SDE_r	FED_r	π_{rr}
1	1.401	0.475	0.127	0.669	0.125	0.857	0.124	0.243
2	0.822	0.762	0.108	0.531	0.062	0.300	0.016	0.208
3	1.677	0.252	0.005	0.835	0.037	1.408	0.070	0.207
4	0.782	1.051	0.042	0.781	0.061	0.511	0.081	0.319
5	1.417	0.397	0.008	1.124	0.329	1.850	0.651	0.333
6	0.914	0.843	0.012	0.732	0.032	0.732	0.052	0.328
7	1.185	0.911	0.435	0.977	0.303	1.042	0.172	0.420
8	0.925	1.071	0.332	0.909	0.186	0.748	0.040	0.410
9	0.952	0.852	0.084	0.917	0.171	0.983	0.258	0.395
10	1.114	1.081	0.574	0.930	0.294	0.779	0.015	0.410
11	1.019	0.851	0.248	0.936	0.249	1.022	0.249	0.392
12	1.791	0.193	0.013	0.975	0.123	1.756	0.233	0.166
13	0.953	0.488	0.001	0.488	0.002	0.488	0.002	0.240
14	1.046	0.507	0.012	0.565	0.025	0.623	0.037	0.263

TABLE 19(S).—Biopterin.

Atom	P_{rr}	SDN_r	FOD_r	SDR_r	FRD_r	SDE_r	FED_r	π_{rr}
1	1.401	0.475	0.128	0.667	0.126	0.860	0.125	0.243
2	0.822	0.762	0.109	0.531	0.063	0.300	0.017	0.208
3	1.677	0.261	0.005	0.835	0.038	1.409	0.071	0.207
4	0.783	1.047	0.038	0.780	0.060	0.513	0.082	0.319
5	1.418	0.396	0.009	1.127	0.332	1.858	0.655	0.332
6	0.917	0.835	0.009	0.729	0.031	0.729	0.053	0.328
7	1.185	0.912	0.437	0.978	0.305	1.043	0.173	0.420
8	0.930	1.062	0.334	0.908	0.189	0.754	0.041	0.410
9	0.945	0.857	0.077	0.922	0.169	0.987	0.260	0.398
10	1.922	1.062	0.561	0.922	0.288	0.782	0.016	0.408
11	1.020	0.850	0.253	0.938	0.253	1.027	0.252	0.392
12	1.792	0.192	0.011	0.970	0.123	1.760	0.235	0.166
13	1.080	0.276	0.000	0.332	0.001	0.387	0.002	0.161
14	0.912	0.371	0.006	0.375	0.013	0.380	0.021	0.172

TABLE 20(P).—β-apo-8,8′-Carotenal.

Atom	$P_{\pi\pi}$	SDN$_r$	FOD$_r$	SDR$_r$	FRD$_r$	SDE$_r$	FED$_r$	$\pi_{\pi\pi}$
1	0.921	2.202	0.200	2.087	0.181	1.972	0.162	0.671
2	1.031	0.832	0.002	0.931	0.007	1.030	0.012	0.409
3	0.968	1.934	0.194	1.819	0.172	1.704	0.150	0.545
4	1.025	1.031	0.013	1.130	0.020	1.228	0.028	0.454
5	0.936	1.759	0.178	1.644	0.155	1.529	0.132	0.485
6	1.034	1.091	0.022	1.286	0.045	1.482	0.068	0.472
7	0.976	1.669	0.159	1.554	0.133	1.439	0.107	0.490
8	1.031	1.190	0.045	1.386	0.069	1.581	0.094	0.483
9	0.942	1.557	0.133	1.442	0.107	1.327	0.080	0.460
10	1.035	1.208	0.058	1.497	0.101	1.787	0.144	0.490
11	0.984	1.493	0.106	1.378	0.079	1.263	0.053	0.476
12	1.024	1.286	0.086	1.576	0.128	1.865	0.170	0.492
13	0.994	1.410	0.078	1.295	0.054	1.180	0.029	0.475
14	0.986	1.364	0.114	1.654	0.152	1.944	0.191	0.482
15	0.979	1.386	0.063	1.181	0.033	0.976	0.003	0.459
16	1.033	1.512	0.143	1.802	0.170	2.092	0.198	0.532
17	0.959	1.225	0.037	1.019	0.018	0.814	0.000	0.424
18	1.038	1.715	0.167	2.005	0.182	2.295	0.195	0.603
19	0.812	0.966	0.025	0.676	0.020	0.387	0.014	0.227
20	1.308	0.881	0.054	0.910	0.053	0.938	0.051	0.271
21	0.953	0.484	0.000	0.484	0.000	0.484	0.000	0.238
22	1.040	0.664	0.024	0.700	0.022	0.736	0.020	0.273
23	0.953	0.484	0.000	0.484	0.000	0.484	0.000	0.238
24	1.040	0.664	0.024	0.700	0.022	0.736	0.020	0.273
25	0.953	0.487	0.000	0.487	0.000	0.487	0.000	0.239
26	1.043	0.616	0.022	0.652	0.019	0.688	0.017	0.267
27	0.953	0.487	0.000	0.487	0.000	0.487	0.000	0.239
28	1.043	0.592	0.016	0.627	0.013	0.663	0.010	0.266
29	0.953	0.487	0.000	0.487	0.000	0.487	0.000	0.239
30	1.049	0.568	0.014	0.654	0.019	0.739	0.024	0.266
31	0.952	0.485	0.000	0.485	0.000	0.485	0.000	0.239
32	1.053	0.608	0.020	0.693	0.023	0.779	0.025	0.270

TABLE 20(S).—β-apo-8,8′-Carotenal.

Atom	P_{rr}	SDN_r	FOD_r	SDR_r	FRD_r	SDE_r	FED_r	π_{rr}
1	0.887	2.143	0.208	2.096	0.183	2.049	0.158	0.682
2	1.053	0.754	0.000	0.942	0.008	1.094	0.016	0.401
3	0.958	1.863	0.205	1.816	0.175	1.769	0.145	0.546
4	1.039	0.956	0.010	1.126	0.020	1.296	0.031	0.449
5	0.934	1.690	0.189	1.643	0.159	1.596	0.128	0.488
6	1.049	0.999	0.016	1.301	0.045	1.603	0.074	0.467
7	0.973	1.596	0.169	1.549	0.137	1.502	0.105	0.491
8	1.043	1.099	0.040	1.401	0.070	1.704	0.099	0.478
9	0.936	1.487	0.140	1.440	0.110	1.392	0.080	0.463
10	1.048	1.102	0.050	1.535	0.102	1.968	0.153	0.485
11	0.988	1.417	0.112	1.370	0.073	1.323	0.054	0.476
12	1.032	1.186	0.081	1.619	0.129	2.052	0.178	0.490
13	1.003	1.328	0.080	1.281	0.056	1.234	0.032	0.473
14	0.984	1.272	0.113	1.705	0.155	2.138	0.197	0.486
15	0.987	1.326	0.069	1.157	0.036	0.988	0.003	0.456
16	1.036	1.418	0.147	1.851	0.176	2.284	0.204	0.534
17	0.967	1.162	0.037	0.993	0.018	0.824	0.000	0.421
18	1.032	1.629	0.175	2.062	0.190	2.495	0.204	0.615
19	0.813	0.956	0.028	0.681	0.022	0.406	0.015	0.228
20	1.308	0.855	0.058	0.924	0.056	0.994	0.053	0.273
21	1.079	0.277	0.000	0.332	0.000	0.388	0.001	0.160
22	0.907	0.459	0.015	0.455	0.013	0.452	0.011	0.175
23	1.079	0.267	0.000	0.332	0.000	0.388	0.001	0.160
24	0.907	0.459	0.015	0.455	0.013	0.452	0.011	0.175
25	1.080	0.277	0.000	0.332	0.000	0.388	0.000	0.161
26	0.910	0.429	0.013	0.426	0.011	0.422	0.009	0.173
27	1.080	0.277	0.000	0.332	0.000	0.388	0.000	0.161
28	0.911	0.415	0.010	0.411	0.008	0.408	0.006	0.173
29	1.080	0.277	0.000	0.333	0.000	0.390	0.001	0.161
30	0.914	0.400	0.008	0.430	0.011	0.461	0.014	0.173
31	1.080	0.277	0.000	0.333	0.000	0.390	0.001	0.161
32	0.916	0.424	0.012	0.455	0.014	0.486	0.015	0.175

TABLE 21(P).—α-Carotene.

Atom	P_{rr}	SDN_r	FOD_r	SDR_r	FRD_r	SDE_r	FED_r	π_{rr}
1	0.994	0.982	0.000	1.083	0.000	1.184	0.000	0.509
2	1.000	0.909	0.000	1.010	0.000	1.111	0.000	0.463
3	0.956	0.472	0.000	0.472	0.000	0.472	0.000	0.229
4	0.929	1.981	0.180	2.310	0.172	2.640	0.164	0.720
5	1.038	0.803	0.002	0.906	0.005	1.010	0.007	0.402
6	0.947	1.656	0.173	1.986	0.165	2.315	0.158	0.533
7	1.039	0.958	0.008	1.171	0.018	1.384	0.027	0.453
8	0.995	1.519	0.159	1.849	0.152	2.179	0.145	0.516
9	1.031	1.088	0.027	1.301	0.034	1.514	0.041	0.471
10	0.966	1.387	0.136	1.717	0.133	2.047	0.131	0.478
11	1.032	1.133	0.042	1.463	0.059	1.792	0.077	0.382
12	1.012	1.315	0.109	1.644	0.111	1.974	0.113	0.492
13	1.016	1.215	0.071	1.544	0.083	1.874	0.095	0.482
14	1.028	1.235	0.079	1.565	0.086	1.894	0.004	0.401
15	0.971	1.281	0.102	1.611	0.108	1.941	0.114	0.466
16	1.027	1.203	0.061	1.417	0.058	1.630	0.056	0.483
17	1.002	1.395	0.133	1.724	0.131	2.054	0.130	0.497
18	1.031	1.102	0.032	1.315	0.036	1.528	0.041	0.473
19	0.960	1.488	0.158	1.818	0.151	2.148	0.144	0.493
20	1.024	1.034	0.018	1.137	0.017	1.241	0.015	0.454
21	0.992	1.663	0.179	1.993	0.166	2.323	0.153	0.553
22	1.031	0.834	0.003	0.937	0.005	1.041	0.008	0.409
23	0.944	1.934	0.188	2.264	0.174	2.593	0.159	0.681
24	1.042	0.689	0.022	0.792	0.021	0.485	0.020	0.272
25	0.953	0.486	0.000	0.486	0.000	0.486	0.000	0.239
26	1.049	0.518	0.000	0.580	0.000	0.642	0.000	0.267
27	0.953	0.486	0.000	0.486	0.000	0.486	0.000	0.239
28	1.050	0.511	0.000	0.573	0.000	0.636	0.000	0.266
29	0.953	0.486	0.000	0.486	0.000	0.486	0.000	0.239
30	1.044	0.602	0.021	0.693	0.020	0.783	0.020	0.268
31	0.953	0.486	0.000	0.486	0.000	0.486	0.000	0.239
32	1.046	0.571	0.017	0.661	0.016	0.752	0.016	0.266
33	0.953	0.487	0.000	0.487	0.000	0.487	0.000	0.239
34	1.047	0.558	0.012	0.648	0.013	0.739	0.014	0.266
35	0.953	0.487	0.000	0.487	0.000	0.487	0.000	0.239

TABLE 21(P).—Continued.

36	1.046	0.583	0.019	0.673	0.019	0.764	0.018	0.267
37	0.953	0.484	0.000	0.484	0.000	0.484	0.000	0.238
38	1.043	0.631	0.023	0.722	0.021	0.812	0.020	0.272
39	0.953	0.484	0.000	0.484	0.000	0.484	0.000	0.238
40	1.043	0.631	0.023	0.722	0.021	0.812	0.020	0.272

TABLE 21(S).—α-Carotene.

Atom	P_{rr}	SDN_r	FOD_r	SDR_r	FRD_r	SDE_r	FED_r	π_{rr}
1	1.016	0.896	0.000	1.072	0.000	1.248	0.000	0.503
2	1.000	0.976	0.000	1.015	0.000	1.154	0.000	0.468
3	1.076	0.271	0.000	0.328	0.000	0.385	0.000	0.156
4	0.913	1.913	0.187	2.452	0.176	2.991	0.164	0.728
5	1.052	0.760	0.002	0.906	0.005	1.052	0.008	0.397
6	0.937	1.584	0.180	2.123	0.169	2.663	0.159	0.538
7	1.054	0.894	0.006	1.205	0.019	2.516	0.031	0.448
8	0.995	1.440	0.166	1.979	0.157	2.518	0.148	0.518
9	1.042	1.028	0.027	1.339	0.035	1.650	0.043	0.467
10	0.963	1.308	0.139	1.847	0.137	2.386	0.135	0.482
11	1.044	1.060	0.040	1.556	0.061	2.052	0.083	0.479
12	1.020	1.224	0.109	1.764	0.114	2.303	0.119	0.492
13	1.022	1.151	0.072	1.647	0.086	2.143	0.099	0.482
14	1.042	1.136	0.075	1.675	0.089	2.214	0.103	0.488
15	0.966	1.228	0.107	1.724	0.111	2.220	0.116	0.471
16	1.039	1.116	0.059	1.470	0.060	1.823	0.060	0.479
17	1.001	1.341	0.143	1.837	0.136	2.333	0.130	0.499
18	1.047	1.011	0.027	1.364	0.037	2.718	0.047	0.468
19	0.950	1.442	0.171	1.938	0.156	2.434	0.142	0.498
20	1.038	0.961	0.016	1.147	0.017	1.333	0.018	0.449
21	0.983	1.617	0.193	2.114	0.172	2.610	0.150	0.557
22	1.053	0.755	0.001	0.941	0.006	1.126	0.011	0.401
23	0.911	1.903	0.199	2.399	0.178	2.895	0.157	0.696
24	0.906	0.479	0.013	0.527	0.013	0.576	0.012	0.186
25	1.079	0.276	0.000	0.332	0.000	0.387	0.000	0.160
26	0.916	0.372	0.000	0.384	0.000	0.397	0.000	0.174
27	1.079	0.276	0.000	0.332	0.000	0.387	0.000	0.160
28	0.915	0.371	0.000	0.381	0.000	0.391	0.000	0.173
29	1.079	0.277	0.000	0.333	0.000	0.390	0.000	0.161
30	0.911	0.471	0.013	0.459	0.012	0.498	0.011	0.174
31	1.080	0.277	0.000	0.333	0.000	0.390	0.000	0.161
32	0.912	0.402	0.010	0.440	0.010	0.479	0.010	0.173
33	1.080	0.277	0.000	0.333	0.000	0.390	0.000	0.161
34	0.913	0.396	0.008	0.432	0.008	0.467	0.008	0.173
35	1.080	0.277	0.000	0.333	0.000	0.390	0.000	0.161
36	0.911	0.411	0.012	0.447	0.011	0.482	0.010	0.173
37	1.079	0.276	0.000	0.333	0.000	0.389	0.000	0.160
38	0.908	0.442	0.014	0.477	0.013	0.512	0.011	0.176
39	1.079	0.276	0.000	0.333	0.000	0.389	0.000	0.160
40	0.908	0.442	0.014	0.477	0.013	0.512	0.011	0.176

TABLE 22 (P).—β-Carotene.

Atom	P_π	SDN_r	FOD_r	SDR_r	FRD_r	SDE_r	FED_r	π_π
1	0.942	1.993	0.168	2.323	0.158	2.652	0.148	0.685
2	1.031	0.834	0.002	0.937	0.004	1.040	0.006	0.409
3	0.990	1.724	0.162	2.054	0.152	2.383	0.142	0.557
4	1.024	1.032	0.013	1.135	0.013	1.239	0.013	0.454
5	0.958	1.550	0.147	1.880	0.141	2.210	0.135	0.497
6	1.031	1.097	0.022	1.311	0.028	1.524	0.034	0.472
7	0.999	1.460	0.129	1.790	0.126	2.120	0.123	0.501
8	1.027	1.195	0.043	1.409	0.045	1.622	0.047	0.482
9	0.968	1.351	0.105	1.681	0.108	2.011	0.111	0.471
10	1.028	1.221	0.056	1.551	0.068	1.880	0.079	0.489
11	1.013	1.292	0.081	1.622	0.088	1.951	0.095	0.488
12	1.013	1.292	0.081	1.622	0.088	1.951	0.095	0.488
13	1.028	1.221	0.056	1.551	0.068	1.880	0.079	0.489
14	0.968	1.351	0.105	1.681	0.108	2.011	0.111	0.471
15	1.027	1.195	0.043	1.409	0.045	1.622	0.047	0.482
16	0.999	1.460	0.120	1.700	0.126	2.120	0.123	0.501
17	1.031	1.097	0.022	1.311	0.028	1.524	0.034	0.472
18	0.958	1.550	0.147	1.880	0.141	2.210	0.135	0.497
19	1.024	1.032	0.013	1.135	0.013	1.239	0.013	0.454
20	0.990	1.724	0.162	2.054	0.152	2.383	0.142	0.557
21	1.031	0.834	0.002	0.937	0.004	1.040	0.006	0.409
22	0.942	1.993	0.168	2.323	0.158	2.652	0.148	0.685
23	0.953	0.484	0.000	0.484	0.000	0.484	0.000	0.238
24	1.043	0.639	0.021	0.729	0.019	0.819	0.018	0.273
25	0.953	0.484	0.000	0.484	0.000	0.484	0.000	0.238
26	1.043	0.639	0.021	0.729	0.019	0.819	0.018	0.273
27	0.953	0.487	0.000	0.487	0.000	0.487	0.000	0.239
28	1.045	0.590	0.018	0.681	0.017	0.771	0.017	0.267
29	0.953	0.487	0.000	0.487	0.000	0.487	0.000	0.239
30	1.047	0.567	0.013	0.657	0.013	0.747	0.014	0.266
31	0.953	0.487	0.000	0.487	0.000	0.487	0.000	0.239
32	1.047	0.567	0.013	0.657	0.013	0.747	0.014	0.266
33	0.953	0.487	0.000	0.487	0.000	0.487	0.000	0.239
34	1.045	0.590	0.018	0.681	0.017	0.771	0.017	0.267
35	0.953	0.484	0.000	0.484	0.000	0.484	0.000	0.238
36	1.043	0.639	0.021	0.729	0.019	0.819	0.018	0.273
37	0.953	0.484	0.000	0.484	0.000	0.484	0.000	0.238
38	1.043	0.639	0.021	0.729	0.019	0.819	0.018	0.273

TABLE 22(S).—β-Carotene.

Atom	P_π	SDN_r	FOD_r	SDR_r	FRD_r	SDE_r	FED_r	π_π
1	0.909	1.940	0.177	2.400	0.161	3.040	0.146	0.700
2	1.053	0.755	0.000	0.943	0.005	1.130	0.009	0.401
3	0.982	1.656	0.173	2.206	0.157	2.755	0.141	0.561
4	1.038	0.960	0.011	1.147	0.013	1.334	0.015	0.449
5	0.948	1.482	0.157	2.032	0.146	2.582	0.136	0.502
6	1.047	1.009	0.018	1.367	0.029	1.725	0.040	0.467
7	0.999	1.385	0.137	1.934	0.131	2.484	0.124	0.503
8	1.039	1.110	0.042	1.469	0.046	1.827	0.050	0.478
9	0.964	1.272	0.109	1.826	0.111	2.376	0.114	0.475
10	1.041	1.126	0.053	1.676	0.070	2.225	0.087	0.486
11	1.020	1.206	0.081	1.755	0.091	2.305	0.100	0.488
12	1.020	1.206	0.081	1.755	0.091	2.305	0.100	0.488
13	1.041	1.126	0.053	1.676	0.070	2.225	0.087	0.486
14	0.964	1.272	0.109	1.826	0.111	2.376	0.114	0.475
15	1.039	1.110	0.042	1.469	0.046	1.827	0.050	0.478
16	0.999	1.385	0.137	1.934	0.131	2.484	0.124	0.503
17	1.047	1.009	0.018	1.367	0.029	1.725	0.040	0.467
18	0.948	1.482	0.157	2.032	0.146	2.582	0.136	0.502
19	1.038	0.960	0.011	1.147	0.013	1.334	0.015	0.449
20	0.982	1.656	0.173	2.206	0.157	2.755	0.141	0.561
21	1.053	0.755	0.000	0.943	0.005	1.130	0.009	0.401
22	0.909	1.940	0.177	2.490	0.161	3.040	0.146	0.700
23	1.079	0.276	0.000	0.333	0.000	0.390	0.000	0.160
24	0.908	0.444	0.012	0.483	0.011	0.523	0.010	0.176
25	1.079	0.276	0.000	0.333	0.000	0.390	0.000	0.160
26	0.908	0.444	0.012	0.483	0.011	0.523	0.010	0.176
27	1.080	0.277	0.000	0.334	0.000	0.390	0.000	0.161
28	0.911	0.414	0.011	0.453	0.010	0.492	0.010	0.173
29	1.080	0.277	0.000	0.333	0.000	0.390	0.000	0.161
30	0.913	0.400	0.008	0.439	0.008	0.478	0.008	0.173
31	1.080	0.277	0.000	0.333	0.000	0.390	0.000	0.161
32	0.913	0.400	0.008	0.439	0.008	0.478	0.008	0.173
33	1.080	0.277	0.000	0.334	0.000	0.390	0.000	0.161
34	0.911	0.414	0.011	0.453	0.010	0.492	0.010	0.173
35	1.079	0.276	0.000	0.333	0.000	0.390	0.000	0.160
36	0.908	0.444	0.012	0.483	0.011	0.523	0.010	0.176
37	1.079	0.276	0.000	0.333	0.000	0.390	0.000	0.160
38	0.908	0.444	0.012	0.483	0.011	0.523	0.010	0.176

TABLE 23.—Chelidanic Acid.

Atom	P_{rr}	SDN_r	FOD_r	SDR_r	FRD_r	SDE_r	FED_r	π_{rr}
1	1.922	0.068	0.014	0.507	0.007	0.946	0.001	0.043
2	0.789	0.862	0.112	0.561	0.057	0.259	0.001	0.202
3	0.888	1.209	0.355	0.970	0.243	0.731	0.131	0.407
4	1.052	1.015	0.094	1.071	0.288	1.127	0.483	0.487
5	0.847	0.779	0.175	0.540	0.099	0.301	0.023	0.208
6	1.052	1.015	0.094	1.071	0.288	1.127	0.483	0.487
7	0.888	1.209	0.355	0.970	0.243	0.731	0.131	0.407
8	0.789	0.862	0.112	0.561	0.057	0.259	0.001	0.202
9	1.922	0.068	0.014	0.507	0.007	0.946	0.001	0.043
10	1.776	0.322	0.173	0.628	0.220	0.935	0.266	0.137
11	1.333	0.628	0.141	0.623	0.083	0.619	0.025	0.248
12	1.409	0.565	0.220	0.745	0.324	0.905	0.428	0.261
13	1.333	0.628	0.141	0.623	0.083	0.619	0.025	0.248

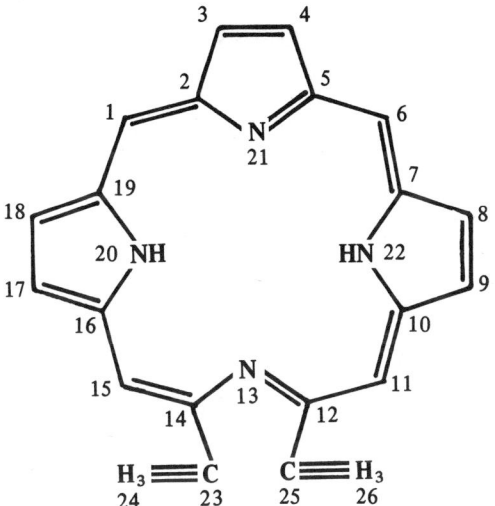

TABLE 24(P).—Chlorin.

Atom	P_{rr}	SDN_r	FOD_r	SDR_r	FRD_r	SDE_r	FED_r	π_{rr}
1	0.950	1.871	0.200	1.625	0.235	1.380	0.270	0.590
2	1.026	0.782	0.017	0.834	0.009	0.886	0.001	0.342
3	1.071	1.041	0.010	1.093	0.006	1.145	0.001	0.455
4	1.071	1.041	0.010	1.093	0.006	1.145	0.001	0.455
5	1.026	0.782	0.017	0.834	0.009	0.886	0.001	0.342
6	0.950	1.871	0.200	1.625	0.235	1.380	0.270	0.590
7	1.000	0.844	0.064	0.896	0.040	0.948	0.017	0.362
8	1.023	1.259	0.135	1.139	0.084	1.020	0.034	0.455
9	1.062	1.038	0.126	1.090	0.080	1.142	0.034	0.445
10	0.957	1.071	0.072	0.951	0.044	0.831	0.017	0.370
11	1.070	1.201	0.176	1.407	0.224	1.609	0.271	0.517
12	0.903	1.154	0.023	1.034	0.012	0.914	0.001	0.413
13	1.243	1.122	0.204	1.319	0.232	1.517	0.260	0.462
14	0.903	1.154	0.023	1.034	0.012	0.914	0.001	0.413
15	1.070	1.201	0.176	1.407	0.224	1.609	0.271	0.517
16	0.957	1.071	0.072	0.951	0.044	0.831	0.017	0.370
17	1.062	1.038	0.126	1.090	0.080	1.142	0.034	0.445
18	1.023	1.259	0.135	1.139	0.084	1.020	0.034	0.455
19	1.000	0.844	0.064	0.896	0.040	0.848	0.017	0.362
20	1.639	0.411	0.000	0.970	0.055	1.530	0.110	0.248
21	1.369	0.823	0.146	1.325	0.187	1.827	0.229	0.435
22	1.639	0.411	0.000	0.970	0.055	1.530	0.110	0.248
23	0.953	0.488	0.000	0.488	0.000	0.488	0.000	0.239
24	1.040	0.543	0.003	0.578	0.002	0.614	0.000	0.265
25	0.953	0.488	0.000	0.488	0.000	0.488	0.000	0.239
26	1.040	0.543	0.003	0.578	0.002	0.614	0.000	0.265

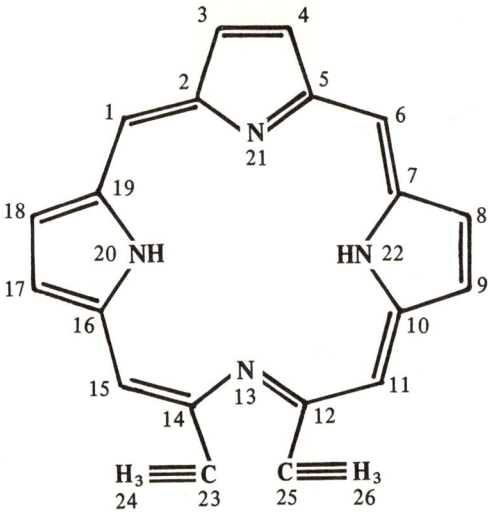

TABLE 24(S).—Chlorin.

Atom	P_π	SDN_r	FOD_r	SDR_r	FRD_r	SDE_r	FED_r	π_π
1	0.950	1.859	0.200	1.620	0.235	1.382	0.271	0.589
2	1.026	0.776	0.017	0.833	0.009	0.891	0.001	0.342
3	1.072	1.034	0.010	1.091	0.006	1.148	0.001	0.455
4	1.072	1.034	0.010	1.091	0.006	1.148	0.001	0.455
5	1.026	0.776	0.017	0.833	0.009	0.891	0.001	0.342
6	0.950	1.859	0.200	1.620	0.235	1.382	0.271	0.589
7	1.002	0.840	0.064	0.897	0.041	0.954	0.017	0.362
8	1.023	1.258	0.135	1.139	0.084	1.020	0.034	0.455
9	1.063	1.034	0.127	1.090	0.080	1.147	0.034	0.455
10	0.956	1.070	0.072	0.951	0.044	0.832	0.017	0.370
11	1.078	1.178	0.178	1.400	0.225	1.622	0.272	0.513
12	0.894	1.156	0.023	1.037	0.012	0.918	0.001	0.416
13	1.249	1.091	0.199	1.304	0.229	1.516	0.259	0.455
14	0.894	1.156	0.023	1.037	0.012	0.918	0.001	0.416
15	1.078	1.178	0.178	1.400	0.225	1.622	0.272	0.513
16	0.956	1.070	0.072	0.951	0.044	0.832	0.017	0.370
17	1.063	1.034	0.127	1.090	0.080	1.147	0.034	0.455
18	1.023	1.258	0.135	1.139	0.084	1.020	0.034	0.455
19	1.002	0.840	0.064	0.897	0.041	0.954	0.017	0.362
20	1.640	0.402	0.000	0.967	0.055	1.532	0.110	0.247
21	1.369	0.822	0.146	1.325	0.187	1.827	0.229	0.425
22	1.640	0.402	0.000	0.967	0.055	1.532	0.110	0.247
23	1.080	0.276	0.000	0.331	0.000	0.387	0.000	0.161
24	0.908	0.391	0.002	0.383	0.001	0.374	0.000	0.173
25	1.080	0.276	0.000	0.331	0.000	0.387	0.000	0.161
26	0.908	0.391	0.002	0.383	0.001	0.374	0.000	0.173

TABLE 25.—Chlorpromazine.

Atom	P_π	SDN_r	FOD_r	SDR_r	FRD_r	SDE_r	FED_r	π_{rr}
1	1.066	0.729	0.250	0.739	0.144	0.750	0.038	0.420
2	1.078	0.729	0.250	0.436	0.179	0.143	0.108	0.440
3	1.044	0.711	0.000	0.722	0.010	0.732	0.021	0.380
4	1.680	0.245	0.000	−3.008	0.536	−6.261	1.071	0.386
5	1.034	0.727	0.000	0.737	0.007	0.748	0.015	0.377
6	1.080	0.726	0.250	0.433	0.180	0.141	0.110	0.441
7	1.050	0.744	0.250	0.755	0.140	0.765	0.029	0.411
8	1.054	0.744	0.000	0.451	0.046	0.158	0.093	0.425
9	1.088	0.717	0.250	0.728	0.149	0.738	0.048	0.428
10	0.985	0.747	0.250	0.454	0.162	0.161	0.074	0.372
11	1.747	0.178	0.000	0.229	0.089	0.281	0.178	0.178
12	0.983	0.748	0.250	0.455	0.161	0.162	0.073	0.371
13	1.102	0.703	0.250	0.797	0.144	0.891	0.038	0.432
14	1.071	0.725	0.000	0.432	0.045	0.140	0.090	0.396
15	1.939	0.051	0.000	0.548	0.008	1.046	0.015	0.040

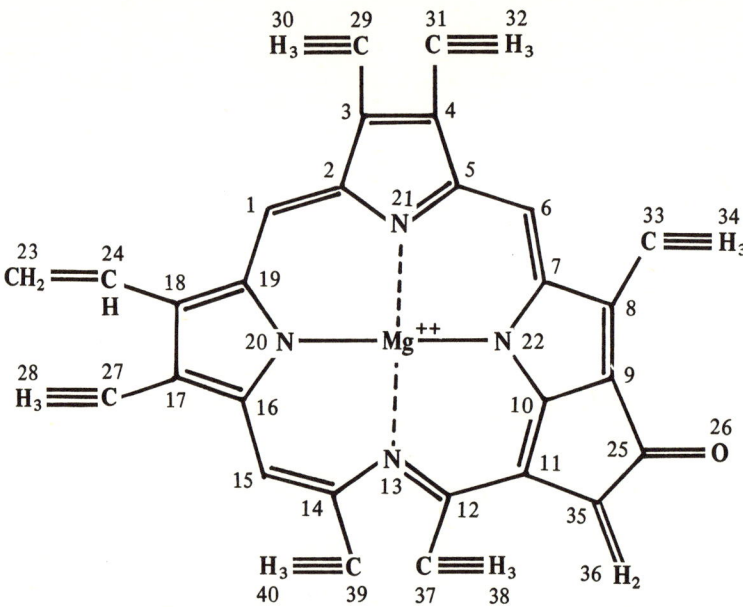

TABLE 26(P).—Chlorophyll *A*.

Atom	P_{rr}	SDN_r	FOD_r	SDR_r	FRD_r	SDE_r	FED_r	π_{rr}
1	0.981	1.721	0.188	1.903	0.228	2.086	0.268	0.631
2	1.032	0.753	0.011	0.839	0.007	0.925	0.003	0.345
3	1.053	0.935	0.012	1.090	0.008	1.245	0.005	0.440
4	1.050	0.952	0.007	1.094	0.006	1.237	0.005	0.442
5	1.036	0.735	0.015	0.833	0.009	0.931	0.002	0.343
6	0.959	1.818	0.166	1.932	0.218	2.046	0.270	0.643
7	1.042	0.743	0.052	0.842	0.027	0.940	0.002	0.341
8	0.969	1.393	0.159	1.235	0.081	1.076	0.003	0.471
9	1.126	0.850	0.091	1.004	0.047	1.158	0.003	0.385
10	1.011	0.855	0.038	0.884	0.021	0.913	0.004	0.358
11	1.041	1.190	0.160	1.729	0.218	2.268	0.275	0.527
12	0.942	0.994	0.016	1.046	0.012	1.097	0.008	0.422
13	1.261	1.103	0.180	1.615	0.214	2.127	0.248	0.482
14	0.936	0.994	0.023	1.020	0.012	1.046	0.001	0.414
15	1.083	1.207	0.156	1.733	0.206	2.259	0.256	0.552
16	1.015	0.854	0.064	0.890	0.034	0.916	0.004	0.351
17	1.037	1.187	0.149	1.273	0.077	1.359	0.004	0.486
18	1.070	0.955	0.105	1.045	0.056	1.135	0.007	0.383
19	1.026	0.792	0.047	0.878	0.024	0.964	0.001	0.354
20	1.402	0.577	0.003	1.500	0.094	2.420	0.185	0.396
21	1.387	0.797	0.124	1.566	0.159	2.335	0.193	0.432
22	1.395	0.563	0.002	1.475	0.105	2.388	0.209	0.391
23	1.041	1.409	0.120	1.499	0.063	1.590	0.007	0.644
24	0.990	0.891	0.008	0.891	0.004	0.891	0.000	0.406
25	0.804	0.889	0.016	0.585	0.008	0.291	0.000	0.214
26	1.310	0.666	0.031	0.657	0.016	0.647	0.001	0.255
27	0.952	0.486	0.000	0.486	0.000	0.486	0.000	0.267
28	1.053	0.545	0.018	0.606	0.009	0.666	0.001	0.267
29	0.952	0.487	0.000	0.487	0.001	0.487	0.000	0.239
30	1.055	0.516	0.001	0.585	0.000	0.654	0.001	0.265
31	0.952	0.487	0.000	0.487	0.001	0.487	0.000	0.239

TABLE 26(P).—*Continued.*

32	1.054	0.518	0.001	0.586	0.000	0.653	0.001	0.265
33	0.952	0.486	0.000	0.486	0.000	0.486	0.000	0.239
34	1.045	0.570	0.020	0.601	0.010	0.632	0.000	0.267
35	0.958	0.480	0.000	0.480	0.000	0.480	0.000	0.233
36	1.033	0.575	0.009	0.662	0.021	0.749	0.033	0.276
37	0.953	0.487	0.000	0.487	0.000	0.487	0.000	0.239
38	1.045	0.524	0.002	0.580	0.001	0.636	0.001	0.265
39	0.953	0.488	0.000	0.488	0.000	0.488	0.000	0.239
40	1.044	0.524	0.003	0.577	0.001	0.630	0.000	0.264

TABLE 26(S).—Chlorophyll *A.*

Atom	P_π	SDN_r	FOD_r	SDR_r	FRD_r	SDE_r	FED_r	π_π
1	0.986	1.662	0.190	1.903	0.231	2.144	0.271	0.628
2	1.035	0.746	0.011	0.841	0.007	0.937	0.002	0.345
3	1.056	0.906	0.013	1.091	0.009	1.277	0.005	0.441
4	1.052	0.923	0.008	1.096	0.007	1.268	0.006	0.444
5	1.040	0.727	0.016	0.835	0.009	0.943	0.002	0.343
6	0.962	1.759	0.162	1.931	0.218	2.103	0.274	0.640
7	1.046	0.733	0.055	0.842	0.028	0.950	0.002	0.340
8	0.960	1.383	0.162	1.236	0.083	1.089	0.003	0.477
9	1.133	0.804	0.086	0.985	0.045	1.166	0.003	0.379
10	1.013	0.838	0.037	0.881	0.021	0.924	0.005	0.357
11	1.041	1.166	0.166	1.751	0.222	2.336	0.279	0.529
12	0.940	0.987	0.017	1.058	0.013	1.129	0.009	0.425
13	1.265	1.066	0.182	1.610	0.215	2.153	0.247	0.475
14	0.930	0.988	0.023	1.026	0.012	1.065	0.000	0.417
15	1.090	1.174	0.163	1.737	0.108	2.300	0.253	0.547
16	1.019	0.854	0.067	0.892	0.036	0.930	0.004	0.351
17	1.030	1.188	0.156	1.283	0.080	1.379	0.004	0.494
18	1.078	0.908	0.101	1.032	0.054	1.156	0.007	0.380
19	1.028	0.785	0.048	0.880	0.025	0.976	0.001	0.354
20	1.405	0.563	0.004	1.510	0.094	2.456	0.184	0.393
21	1.394	0.762	0.122	1.572	0.157	2.381	0.193	0.427
22	1.401	0.550	0.004	1.498	0.108	2.445	0.212	0.388
23	1.046	1.364	0.117	1.488	0.062	1.612	0.008	0.641
24	0.990	0.892	0.008	0.892	0.004	0.892	0.000	0.406
25	0.801	0.878	0.017	0.580	0.009	0.281	0.000	0.213
26	1.315	0.642	0.032	0.647	0.017	0.652	0.002	0.252
27	1.079	0.277	0.000	0.332	0.000	0.388	0.000	0.161
28	0.916	0.393	0.011	0.400	0.006	0.407	0.000	0.174
29	1.079	0.276	0.000	0.332	0.000	0.388	0.000	0.161
30	0.908	0.374	0.001	0.387	0.001	0.400	0.000	0.173
31	1.079	0.276	0.000	0.332	0.000	0.388	0.000	0.161
32	0.908	0.374	0.001	0.387	0.001	0.400	0.000	0.173
33	1.079	0.277	0.000	0.332	0.000	0.387	0.000	0.161
34	0.912	0.407	0.011	0.396	0.006	0.386	0.000	0.173
35	1.079	0.272	0.000	0.328	0.000	0.384	0.001	0.157
36	0.900	0.412	0.005	0.436	0.012	0.459	0.019	0.179
37	1.080	0.276	0.000	0.332	0.000	0.387	0.000	0.161
38	0.911	0.380	0.001	0.384	0.001	0.389	0.001	0.173
39	1.080	0.276	0.000	0.332	0.000	0.387	0.000	0.161
40	0.911	0.380	0.001	0.384	0.001	0.389	0.001	0.173

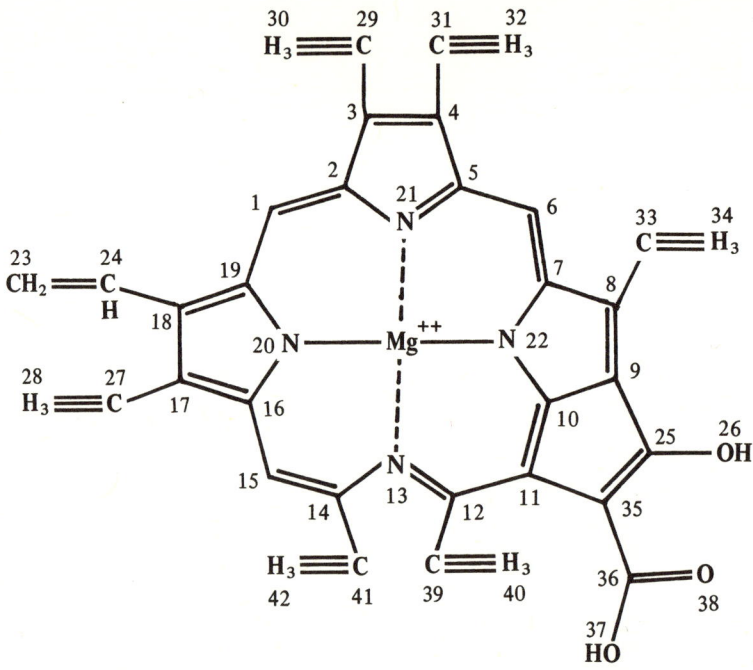

TABLE 27(P).—Chlorophyll *A* (Tautomer 1).

Atom	P_{rr}	SDN_r	FOD_r	SDR_r	FRD_r	SDE_r	FED_r	π_{rr}
1	0.931	6.777	0.186	3.845	0.141	0.912	0.095	0.524
2	0.919	2.696	0.057	1.754	0.062	0.813	0.066	0.389
3	0.989	2.457	0.049	1.731	0.083	1.006	0.118	0.439
4	0.959	3.028	0.064	1.957	0.034	0.886	0.003	0.438
5	0.951	2.119	0.042	1.520	0.092	0.921	0.142	0.389
6	0.868	7.934	0.216	4.345	0.124	0.756	0.033	0.507
7	0.977	2.157	0.048	1.558	0.069	0.958	0.089	0.382
8	0.866	7.655	0.218	4.245	0.125	0.834	0.032	0.494
9	1.107	1.612	0.036	1.287	0.037	0.962	0.038	0.338
10	0.957	2.374	0.041	1.756	0.147	1.139	0.253	0.464
11	1.109	1.814	0.042	1.450	0.033	1.085	0.023	0.357
12	0.688	2.027	0.002	1.349	0.005	0.671	0.007	0.432
13	1.699	1.567	0.041	1.201	0.041	0.836	0.041	0.167
14	0.758	5.420	0.138	3.120	0.150	0.820	0.162	0.470
15	1.085	2.319	0.051	1.776	0.122	1.233	0.192	0.459
16	0.861	5.268	0.143	2.969	0.095	0.969	0.046	0.386
17	1.013	3.086	0.076	2.144	0.126	1.203	0.175	0.469
18	0.976	4.735	0.131	2.746	0.066	0.758	0.001	0.383
19	0.925	2.803	0.064	2.861	0.108	0.920	0.152	0.406
20	1.759	0.401	0.004	0.641	0.008	0.880	0.012	0.130
21	1.750	1.621	0.048	1.233	0.027	0.844	0.006	0.137
22	1.767	0.290	0.000	0.625	0.009	0.959	0.017	0.128
23	0.956	5.184	0.131	3.195	0.066	1.206	0.001	0.657
24	0.994	0.887	0.000	0.887	0.000	0.887	0.000	0.407
25	0.976	3.033	0.067	2.115	0.062	1.196	0.058	0.505
26	1.960	0.223	0.006	0.640	0.007	1.058	0.008	0.029
27	0.952	0.487	0.000	0.487	0.000	0.487	0.000	0.239
28	1.050	0.779	0.009	0.713	0.017	0.648	0.025	0.266
29	0.952	0.487	0.000	0.487	0.000	0.487	0.000	0.239

TABLE 27(P).—*Continued.*

30	1.047	0.702	0.006	0.663	0.011	0.625	0.017	0.265
31	0.952	0.487	0.000	0.487	0.000	0.487	0.000	0.239
32	1.044	0.772	0.002	0.691	0.004	0.609	0.000	0.266
33	0.952	0.486	0.000	0.486	0.000	0.486	0.000	0.239
34	1.033	1.335	0.027	0.967	0.016	0.599	0.005	0.269
35	1.153	1.324	0.023	1.357	0.074	1.390	0.124	0.392
36	0.799	0.886	0.002	0.587	0.003	0.288	0.004	0.213
37	1.966	0.032	0.000	0.505	0.000	0.979	0.001	0.020
38	1.344	0.705	0.006	0.708	0.016	0.711	0.026	0.245
39	0.955	0.486	0.000	0.486	0.000	0.486	0.000	0.239
40	1.019	0.644	0.000	0.610	0.001	0.577	0.001	0.269
41	0.955	0.486	0.001	0.486	0.001	0.486	0.000	0.239
42	1.026	0.661	0.017	0.829	0.020	0.597	0.023	0.269

TABLE 27(S).—Chlorophyll *A* (Tautomer 1).

Atom	P_{rr}	SDN_r	FOD_r	SDR_r	FRD_r	SDE_r	FED_r	π_{rr}
1	0.933	5.447	0.190	3.183	0.144	0.920	0.098	0.523
2	0.924	2.224	0.055	1.523	0.063	0.823	0.070	0.389
3	0.989	2.147	0.053	1.585	0.088	1.022	0.122	0.441
4	0.957	2.558	0.066	1.728	0.034	0.897	0.003	0.440
5	0.957	1.809	0.043	1.372	0.095	0.934	0.074	0.388
6	0.868	6.285	0.215	3.521	0.124	0.757	0.033	0.505
7	0.983	1.847	0.051	1.410	0.072	0.792	0.093	0.382
8	0.858	6.067	0.219	3.454	0.127	0.842	0.036	0.497
9	1.111	1.287	0.034	1.126	0.036	0.965	0.037	0.335
10	0.958	2.019	0.039	1.581	0.144	1.144	0.249	0.462
11	1.114	1.491	0.043	1.292	0.032	1.094	0.020	0.353
12	0.676	1.970	0.002	1.319	0.005	0.669	0.009	0.429
13	1.699	1.250	0.041	1.041	0.041	0.833	0.040	0.166
14	0.748	4.456	0.141	2.643	0.155	0.829	0.169	0.471
15	1.092	2.008	0.056	1.630	0.127	1.253	0.198	0.457
16	0.865	4.304	0.149	2.490	0.099	0.676	0.049	0.387
17	1.007	2.625	0.081	1.924	0.132	1.223	0.183	0.476
18	0.981	3.730	0.130	2.247	0.065	0.765	0.001	0.381
19	0.927	2.333	0.065	1.632	0.111	0.931	0.158	0.406
20	1.760	0.371	0.004	0.627	0.008	0.882	0.012	0.129
21	1.725	1.245	0.047	1.046	0.026	0.847	0.006	0.136
22	1.769	0.279	0.000	0.621	0.008	0.964	0.015	0.126
23	0.960	4.182	0.130	2.699	0.066	1.216	0.002	0.565
24	0.994	0.888	0.000	0.888	0.000	0.888	0.000	0.408
25	0.978	2.510	0.068	1.871	0.060	1.203	0.051	0.503
26	1.960	0.179	0.006	0.619	0.007	1.059	0.007	0.029
27	1.079	0.279	0.000	0.334	0.001	0.388	0.001	0.161
28	0.915	0.496	0.006	0.446	0.010	0.396	0.014	0.173
29	1.079	0.279	0.000	0.334	0.001	0.388	0.001	0.161
30	0.914	0.462	0.004	0.422	0.007	0.382	0.010	0.173
31	1.079	0.279	0.000	0.334	0.001	0.388	0.001	0.161
32	0.911	0.491	0.005	0.432	0.002	0.373	0.000	0.173
33	1.079	0.279	0.000	0.334	0.001	0.388	0.001	0.161
34	0.904	0.739	0.015	0.553	0.009	0.367	0.003	0.174
35	1.154	1.125	0.022	1.259	0.069	1.393	0.116	0.390
36	0.799	0.868	0.002	0.578	0.003	0.288	0.004	0.213
37	1.966	0.031	0.000	0.505	0.000	0.979	0.001	0.020
38	1.344	0.654	0.006	0.683	0.015	0.712	0.025	0.245
39	1.080	0.276	0.000	0.330	0.000	0.384	0.000	0.160
40	0.894	0.446	0.000	0.400	0.000	0.354	0.001	0.173
41	1.080	0.281	0.000	0.333	0.001	0.385	0.001	0.160
42	0.899	0.624	0.010	0.495	0.012	0.366	0.013	0.173

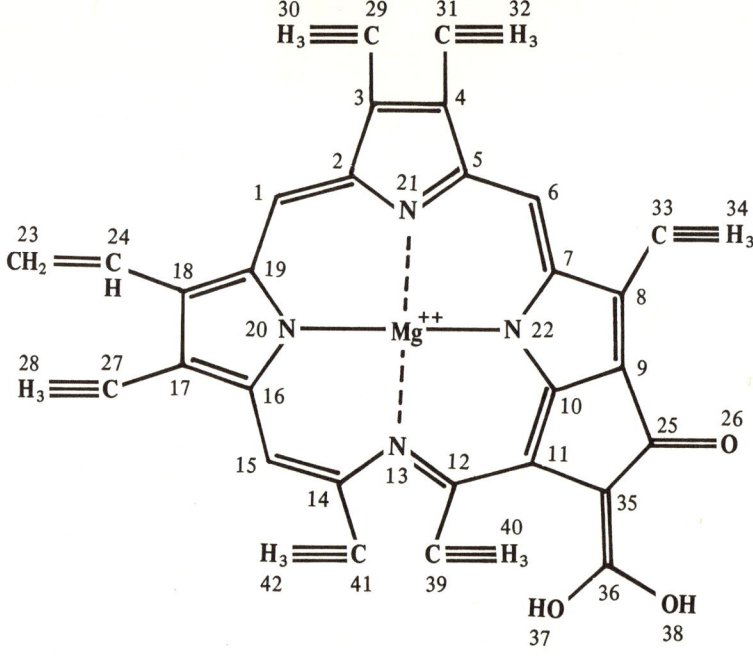

TABLE 28(P).—Chlorophyll *A* (Tautomer 2).

Atom	P_π	SDN_r	FOD_r	SDR_r	FRD_r	SDE_r	FED_r	π_{rr}
1	0.950	3.078	0.175	2.028	0.102	0.979	0.029	0.524
2	0.933	1.448	0.044	1.162	0.062	0.876	0.081	0.389
3	0.991	1.553	0.052	1.279	0.026	1.006	0.000	0.437
4	0.978	1.574	0.044	1.273	0.061	0.792	0.077	0.438
5	0.947	1.429	0.051	1.164	0.038	0.906	0.025	0.988
6	0.920	3.150	0.160	2.048	0.124	0.946	0.088	0.530
7	0.965	1.463	0.068	1.204	0.070	0.945	0.072	0.387
8	0.900	2.676	0.146	1.749	0.073	0.822	0.000	0.451
9	1.098	1.345	0.068	1.195	0.071	1.045	0.074	0.384
10	0.887	1.862	0.063	1.345	0.032	0.828	0.002	0.418
11	1.068	1.510	0.082	1.330	0.154	1.150	0.226	0.392
12	0.763	2.307	0.074	1.675	0.182	1.042	0.290	0.512
13	1.695	1.151	0.081	1.001	0.060	0.851	0.038	0.174
14	0.782	2.421	0.105	1.659	0.077	0.897	0.049	0.470
15	1.080	1.610	0.082	1.415	0.044	1.221	0.006	0.463
16	0.884	2.276	0.121	1.513	0.092	0.751	0.063	0.391
17	1.013	1.771	0.093	1.485	0.047	1.199	0.001	0.467
18	0.992	2.115	0.121	1.469	0.088	0.823	0.055	0.383
19	0.933	1.524	0.060	1.237	0.055	0.951	0.049	0.403
20	1.760	0.313	0.002	0.598	0.001	0.883	0.000	0.129
21	1.755	0.660	0.044	0.760	0.025	0.861	0.006	0.133
22	1.755	0.297	0.000	0.593	0.018	0.889	0.036	0.132
23	0.972	2.570	0.123	1.924	0.098	1.278	0.074	0.653
24	0.995	0.890	0.001	0.890	0.006	0.890	0.010	0.408
25	0.829	0.953	0.012	0.635	0.015	0.316	0.018	0.212
26	1.337	0.925	0.028	0.873	0.067	0.820	0.105	0.277
27	0.952	0.487	0.000	0.487	0.000	0.487	0.000	0.239
28	1.050	0.618	0.011	0.632	0.006	0.647	0.000	0.266
29	0.952	0.487	0.000	0.487	0.000	0.487	0.000	0.239

TABLE 28(P).—*Continued.*

30	1.048	0.592	0.006	0.608	0.003	0.625	0.000	0.265
31	0.952	0.487	0.000	0.487	0.000	0.487	0.000	0.239
32	1.046	0.594	0.005	0.607	0.008	0.620	0.011	0.265
33	0.952	0.487	0.000	0.487	0.000	0.487	0.000	0.239
34	1.037	0.727	0.018	0.664	0.009	0.600	0.000	0.267
35	1.132	0.636	0.002	0.960	0.081	1.283	0.160	0.367
36	0.833	1.713	0.030	1.430	0.136	1.148	0.241	0.549
37	1.948	0.092	0.003	0.566	0.018	1.041	0.033	0.037
38	1.948	0.092	0.003	0.566	0.018	1.041	0.033	0.037
39	0.955	0.486	0.000	0.486	0.000	0.486	0.000	0.239
40	1.027	0.678	0.009	0.651	0.024	0.623	0.039	0.270
41	0.955	0.486	0.000	0.486	0.000	0.486	0.000	0.239
42	1.029	0.694	0.013	0.651	0.010	0.607	0.007	0.268

TABLE 28(S).—Chlorophyll *A* (Tautomer 2).

Atom	P_{rr}	SDN_r	FOD_r	SDR_r	FRD_r	SDE_r	FED_r	π_{rr}
1	0.952	2.873	0.181	1.929	0.104	0.985	0.027	0.523
2	0.938	1.362	0.042	1.125	0.063	0.888	0.084	0.389
3	0.990	1.495	0.056	1.258	0.028	1.022	0.000	0.439
4	0.977	1.495	0.045	1.240	0.062	0.986	0.079	0.440
5	0.953	1.357	0.053	1.138	0.041	0.919	0.028	0.388
6	0.920	2.905	0.157	1.927	0.122	0.949	0.086	0.528
7	0.971	1.399	0.072	1.180	0.074	0.960	0.076	0.387
8	0.892	2.491	0.145	1.659	0.072	0.827	0.000	0.454
9	1.105	1.229	0.066	1.142	0.071	1.055	0.076	0.380
10	0.888	1.757	0.061	1.294	0.031	0.831	0.001	0.417
11	1.076	1.407	0.085	1.290	0.157	1.172	0.230	0.390
12	0.751	2.172	0.070	1.610	0.181	1.047	0.293	0.509
13	1.694	1.031	0.080	0.939	0.058	0.848	0.036	0.172
14	0.773	2.307	0.108	1.608	0.081	0.909	0.053	0.472
15	1.088	1.517	0.086	1.378	0.045	1.240	0.005	0.460
16	0.887	2.158	0.128	1.459	0.097	0.761	0.065	0.392
17	1.008	1.694	0.098	1.457	0.050	1.220	0.002	0.473
18	0.936	1.439	0.060	1.202	0.056	0.965	0.053	0.404
19	0.998	1.940	0.121	1.385	0.087	0.830	0.054	0.381
20	1.761	0.306	0.003	0.595	0.002	0.884	0.000	0.128
21	1.757	0.596	0.043	0.729	0.024	0.863	0.006	0.132
22	1.756	0.287	0.000	0.590	0.018	0.893	0.036	0.131
23	0.977	2.397	0.123	1.842	0.098	1.287	0.073	0.651
24	0.994	0.891	0.001	0.891	0.006	0.891	0.010	0.408
25	0.829	0.936	0.012	0.627	0.015	0.317	0.018	0.213
26	1.339	0.881	0.028	0.854	0.067	0.827	0.106	0.276
27	1.079	0.278	0.000	0.333	0.000	0.388	0.000	0.161
28	0.915	0.430	0.007	0.413	0.004	0.396	0.000	0.173
29	1.079	0.277	0.000	0.332	0.000	0.387	0.000	0.161
30	0.914	0.416	0.004	0.399	0.002	0.382	0.000	0.173
31	1.079	0.277	0.000	0.332	0.000	0.387	0.000	0.161
32	0.913	0.415	0.003	0.397	0.004	0.397	0.006	0.173
33	1.079	0.278	0.000	0.332	0.000	0.387	0.000	0.161
34	0.907	0.485	0.010	0.426	0.005	0.367	0.000	0.173
35	1.131	0.634	0.002	0.961	0.080	1.287	0.159	0.368
36	0.835	1.677	0.032	1.420	0.137	1.162	0.242	0.551
37	1.948	0.089	0.003	0.565	0.018	1.042	0.033	0.037
38	1.948	0.089	0.003	0.565	0.018	1.042	0.033	0.037
39	1.080	0.276	0.000	0.331	0.001	0.385	0.002	0.160
40	0.899	0.461	0.005	0.421	0.014	0.381	0.022	0.174
41	1.080	0.277	0.000	0.331	0.000	0.385	0.000	0.160
42	0.901	0.471	0.008	0.421	0.006	0.372	0.004	0.173

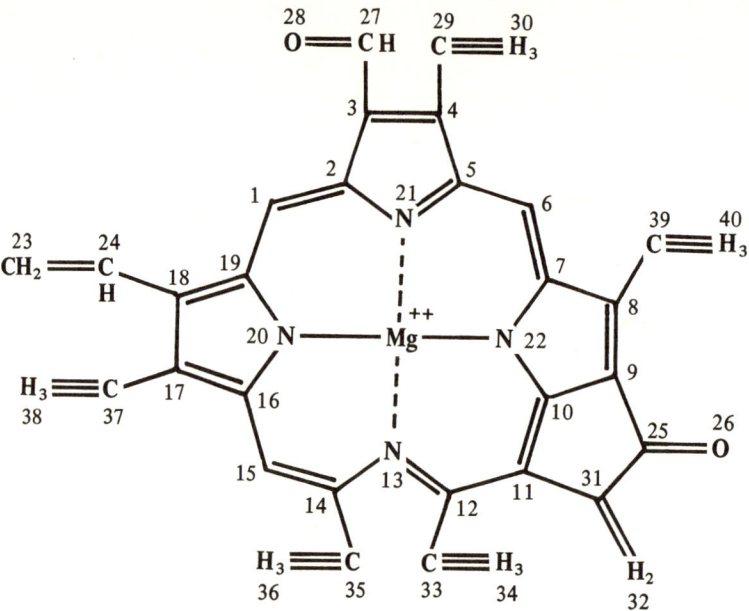

TABLE 29(P).—Chlorophyll *B*..

Atom	P_π	SDN_r	FOD_r	SDR_r	FRD_r	SDE_r	FED_r	π_π
1	0.989	1.784	0.222	1.949	0.247	2.113	0.272	0.641
2	1.015	0.745	0.011	0.831	0.007	0.897	0.004	0.347
3	1.117	0.846	0.017	0.991	0.010	1.137	0.004	0.387
4	0.971	1.274	0.000	1.170	0.002	1.066	0.003	0.466
5	1.038	0.721	0.018	0.817	0.010	0.913	0.002	0.339
6	0.947	1.903	0.119	1.948	0.191	1.992	0.263	0.649
7	1.041	0.741	0.049	0.836	0.025	0.932	0.002	0.341
8	0.966	1.407	0.133	1.235	0.068	1.062	0.003	0.471
9	1.125	0.851	0.082	1.003	0.043	1.155	0.004	0.385
10	1.009	0.864	0.030	0.882	0.017	0.900	0.003	0.358
11	1.040	1.204	0.163	1.731	0.219	2.257	0.275	0.528
12	0.941	1.004	0.011	1.047	0.010	1.089	0.009	0.422
13	1.259	1.116	0.178	1.613	0.213	2.111	0.247	0.583
14	0.933	1.019	0.029	1.024	0.015	1.030	0.000	0.414
15	1.081	1.218	0.153	1.726	0.203	2.234	0.252	0.552
16	1.012	0.895	0.073	0.900	0.039	0.906	0.005	0.352
17	1.034	1.216	0.158	1.282	0.081	1.348	0.003	0.487
18	1.068	0.985	0.117	1.058	0.062	1.132	0.007	0.385
19	1.021	0.811	0.049	0.877	0.025	0.943	0.001	0.355
20	1.404	0.579	0.006	1.501	0.095	2.423	0.185	0.394
21	1.368	0.855	0.133	1.569	0.169	2.284	0.204	0.443
22	1.393	0.589	0.006	1.479	0.106	2.370	0.207	0.396
23	1.039	1.438	0.133	1.512	0.070	1.568	0.008	0.646
24	0.990	0.891	0.008	0.891	0.004	0.891	0.000	0.406
25	0.804	0.889	0.015	0.585	0.007	0.281	0.000	0.214
26	1.310	0.667	0.028	0.657	0.014	0.646	0.001	0.255
27	0.796	0.892	0.003	0.587	0.002	0.283	0.000	0.218
28	1.303	0.650	0.006	0.638	0.004	0.626	0.001	0.250
29	0.952	0.487	0.000	0.487	0.000	0.487	0.000	0.239
30	1.046	0.556	0.000	0.593	0.000	0.631	0.000	0.266

TABLE 29(P).—*Continued.*

31	0.958	0.480	0.000	0.480	0.000	0.480	0.000	0.233
32	1.032	0.576	0.010	0.662	0.021	0.748	0.032	0.276
33	0.953	0.487	0.000	0.487	0.000	0.487	0.000	0.239
34	1.045	0.525	0.001	0.580	0.001	0.635	0.001	0.265
35	0.953	0.488	0.000	0.488	0.000	0.488	0.000	0.239
36	1.044	0.527	0.004	0.578	0.002	0.628	0.000	0.264
37	0.952	0.486	0.000	0.486	0.000	0.486	0.000	0.239
38	1.053	0.549	0.020	0.627	0.010	0.665	0.000	0.267
39	0.952	0.486	0.000	0.486	0.000	0.486	0.000	0.239
40	1.045	0.572	0.016	0.601	0.008	0.630	0.000	0.267

TABLE 29(S).—Chlorophyll *B*.

Atom	$P_{\pi\pi}$	SDN_r	FOD_r	SDR_r	FRD_r	SDE_r	FED_r	π_{rr}
1	0.995	1.722	0.226	1.947	0.250	2.173	0.275	0.638
2	1.016	0.760	0.010	0.835	0.007	0.907	0.004	0.347
3	1.124	0.805	0.019	0.976	0.011	1.147	0.004	0.382
4	0.983	1.266	0.000	1.173	0.002	1.079	0.003	0.471
5	1.042	0.712	0.019	0.817	0.011	0.923	0.002	0.339
6	0.949	1.846	0.111	1.944	0.189	2.042	0.266	0.647
7	1.045	0.730	0.051	0.836	0.026	0.942	0.002	0.340
8	0.957	1.397	0.132	1.235	0.067	1.074	0.003	0.477
9	1.133	0.804	0.077	0.984	0.040	1.163	0.004	0.379
10	1.011	0.848	0.028	0.879	0.016	0.909	0.004	0.357
11	1.040	1.180	0.168	1.752	0.223	2.323	0.279	0.530
12	0.938	0.997	0.011	1.058	0.011	1.120	0.011	0.425
13	1.263	1.081	0.180	1.607	0.213	2.134	0.245	0.476
14	0.927	1.041	0.030	1.031	0.015	1.047	0.000	0.417
15	1.088	1.188	0.160	1.728	0.204	2.268	0.249	0.548
16	1.015	0.886	0.077	0.902	0.041	0.919	0.005	0.352
17	1.027	1.220	0.166	1.293	0.085	1.367	0.003	0.496
18	1.076	0.937	0.113	1.045	0.061	1.153	0.008	0.381
19	1.022	0.806	0.051	0.880	0.026	0.953	0.001	0.355
20	1.408	0.565	0.006	1.511	0.095	2.457	0.184	0.391
21	1.372	0.827	0.132	1.576	0.169	2.326	0.205	0.440
22	1.398	0.576	0.008	1.501	0.109	2.425	0.210	0.392
23	1.045	1.393	0.130	1.501	0.070	1.608	0.009	0.643
24	0.990	0.891	0.004	0.891	0.004	0.891	0.000	0.406
25	0.801	0.878	0.015	0.579	0.008	0.281	0.000	0.213
26	1.315	0.642	0.029	0.647	0.015	0.652	0.002	0.252
27	0.796	0.887	0.003	0.585	0.002	0.283	0.000	0.218
28	1.304	0.638	0.007	0.633	0.004	0.628	0.001	0.250
29	1.079	0.276	0.000	0.332	0.000	0.387	0.000	0.161
30	0.912	0.399	0.000	0.392	0.000	0.385	0.000	0.173
31	1.079	0.272	0.000	0.328	0.000	0.384	0.001	0.157
32	0.900	0.413	0.006	0.436	0.012	0.458	0.018	0.178
33	1.080	0.276	0.000	0.332	0.000	0.387	0.000	0.161
34	0.911	0.380	0.001	0.384	0.001	0.389	0.000	0.173
35	1.080	0.276	0.000	0.332	0.000	0.387	0.000	0.161
36	0.911	0.381	0.002	0.383	0.001	0.384	0.000	0.173
37	1.079	0.277	0.000	0.332	0.000	0.388	0.000	0.161
38	0.916	0.396	0.012	0.401	0.006	0.406	0.000	0.174
39	1.079	0.277	0.000	0.332	0.000	0.387	0.000	0.161
40	0.912	0.408	0.009	0.396	0.005	0.385	0.000	0.173

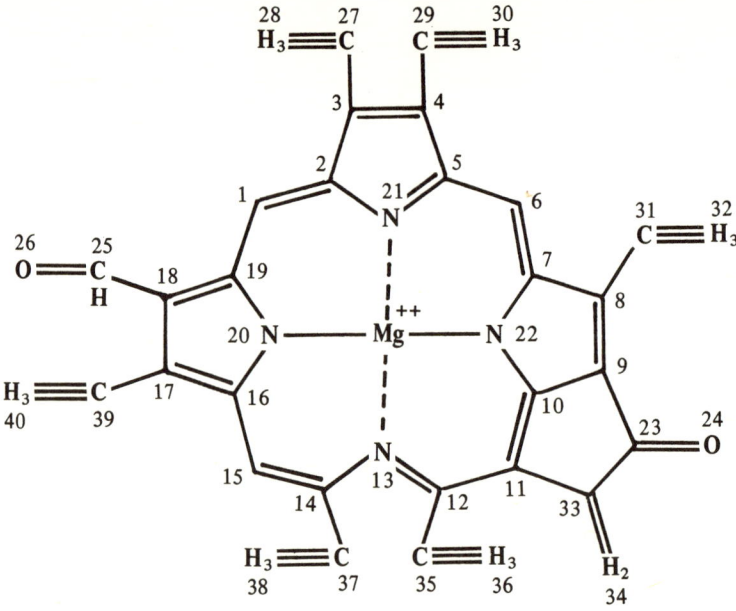

TABLE 30(P).—Chlorophyll *D*.

Atom	P_{rr}	SDN_r	FOD_r	SDR_r	FRD_r	SDE_r	FED_r	π_{rr}
1	0.979	1.813	0.210	1.963	0.244	2.113	0.278	0.648
2	1.030	0.753	0.008	0.830	0.005	0.906	0.002	0.344
3	1.051	0.955	0.016	1.097	0.011	1.239	0.006	0.441
4	1.048	0.955	0.005	1.091	0.005	1.226	0.005	0.442
5	1.034	0.752	0.019	0.835	0.011	0.918	0.003	0.343
6	0.960	1.815	0.151	1.929	0.211	2.043	0.271	0.643
7	1.040	0.760	0.054	0.843	0.028	0.926	0.001	0.342
8	0.967	1.411	0.155	1.241	0.079	1.071	0.004	0.472
9	1.124	0.867	0.091	1.008	0.047	1.149	0.003	0.386
10	0.861	2.093	0.065	1.394	0.077	0.695	0.090	0.388
11	1.035	1.244	0.169	1.725	0.218	2.207	0.267	0.531
12	0.941	0.998	0.016	1.041	0.011	1.083	0.007	0.421
13	1.254	1.160	0.188	1.619	0.215	2.078	0.242	0.487
14	0.937	0.993	0.024	1.019	0.012	1.045	0.001	0.414
15	1.074	1.267	0.165	1.741	0.208	2.214	0.251	0.558
16	1.020	0.862	0.065	0.888	0.034	0.914	0.002	0.350
17	0.983	1.357	0.180	1.231	0.091	1.105	0.002	0.469
18	1.111	0.960	0.108	1.047	0.056	1.135	0.005	0.394
19	1.022	0.773	0.041	0.849	0.022	0.926	0.002	0.347
20	1.393	0.600	0.006	1.469	0.105	2.337	0.204	0.398
21	1.388	0.816	0.125	1.580	0.161	2.345	0.197	0.435
22	1.394	0.571	0.004	1.467	0.104	2.263	0.205	0.396
23	0.803	0.890	0.016	0.586	0.008	0.281	0.000	0.214
24	1.309	0.671	0.031	0.658	0.016	0.645	0.001	0.255
25	0.796	0.904	0.020	0.594	0.010	0.283	0.000	0.218
26	1.302	0.683	0.038	0.655	0.020	0.626	0.001	0.251
27	0.952	0.487	0.000	0.487	0.000	0.487	0.000	0.239
28	1.054	0.518	0.002	0.586	0.001	0.653	0.001	0.265
29	0.952	0.487	0.000	0.487	0.000	0.487	0.000	0.239
30	1.054	0.518	0.001	0.585	0.001	0.652	0.001	0.265

TABLE 30(P).—*Continued.*

31	0.952	0.486	0.000	0.486	0.000	0.486	0.000	0.239
32	1.045	0.573	0.019	0.602	0.010	0.631	0.000	0.267
33	0.958	0.480	0.000	0.480	0.000	0.480	0.000	0.233
34	1.032	0.579	0.010	0.661	0.021	0.743	0.032	0.276
35	0.953	0.487	0.000	0.487	0.000	0.487	0.000	0.239
36	1.045	0.524	0.002	0.579	0.001	0.634	0.001	0.265
37	0.953	0.488	0.000	0.488	0.000	0.488	0.000	0.239
38	1.044	0.524	0.003	0.577	0.002	0.630	0.000	0.264
39	0.952	0.487	0.000	0.487	0.000	0.487	0.000	0.239
40	1.047	0.566	0.022	0.601	0.011	0.635	0.000	0.266

TABLE 30(S).—Chlorophyll *D.*

Atom	P_{rr}	SDN_r	FOD_r	SDR_r	FRD_r	SDE_r	FED_r	π_{rr}
1	0.984	1.742	0.210	1.957	0.246	2.172	0.282	0.644
2	1.033	0.746	0.008	0.832	0.005	0.917	0.002	0.344
3	1.054	0.924	0.017	1.098	0.012	1.272	0.006	0.443
4	1.051	0.926	0.005	1.092	0.005	1.257	0.005	0.444
5	1.037	0.743	0.020	0.837	0.011	0.931	0.003	0.343
6	0.963	1.756	0.149	1.927	0.212	2.098	0.275	0.640
7	1.044	0.749	0.057	0.842	0.029	0.936	0.001	0.341
8	0.959	1.399	0.158	1.241	0.081	1.084	0.004	0.477
9	1.132	0.818	0.087	0.988	0.045	1.157	0.003	0.379
10	0.864	1.898	0.061	1.298	0.076	0.698	0.090	0.386
11	1.034	1.215	0.174	1.743	0.222	2.271	0.270	0.533
12	0.939	0.991	0.017	1.052	0.013	1.114	0.008	0.425
13	1.258	1.117	0.190	1.610	0.216	2.102	0.241	0.480
14	0.931	0.987	0.024	1.026	0.012	1.064	0.001	0.417
15	1.082	1.229	0.171	1.741	0.210	2.253	0.249	0.553
16	1.024	0.852	0.068	0.890	0.036	0.928	0.004	0.350
17	0.976	1.345	0.185	1.233	0.093	1.121	0.001	0.475
18	1.119	0.909	0.103	1.028	0.054	1.147	0.005	0.389
19	1.024	0.767	0.043	0.853	0.022	0.938	0.002	0.348
20	1.396	0.586	0.007	1.479	0.105	2.372	0.204	0.395
21	1.395	0.777	0.123	1.584	0.160	2.391	0.197	0.429
22	1.399	0.558	0.005	1.488	0.107	2.417	0.208	0.390
23	0.801	0.880	0.017	0.580	0.009	0.281	0.000	0.213
24	1.315	0.649	0.032	0.648	0.017	0.649	0.001	0.252
25	0.796	0.898	0.020	0.591	0.010	0.284	0.000	0.218
26	1.303	0.668	0.037	0.648	0.019	0.629	0.001	0.250
27	1.079	0.276	0.000	0.332	0.000	0.388	0.000	0.161
28	0.918	0.375	0.001	0.388	0.001	0.400	0.000	0.173
29	1.079	0.276	0.000	0.332	0.000	0.388	0.000	0.161
30	0.918	0.375	0.000	0.387	0.000	0.399	0.000	0.173
31	1.079	0.277	0.000	0.332	0.000	0.387	0.000	0.161
32	0.912	0.408	0.011	0.397	0.006	0.386	0.000	0.173
33	1.079	0.272	0.000	0.328	0.000	0.384	0.001	0.157
34	0.900	0.414	0.006	0.435	0.012	0.955	0.018	0.178
35	1.080	0.276	0.000	0.332	0.000	0.387	0.000	0.161
36	0.911	0.380	0.001	0.384	0.001	0.388	0.000	0.173
37	1.080	0.276	0.000	0.332	0.000	0.387	0.000	0.161
38	0.911	0.379	0.002	0.382	0.001	0.385	0.000	0.173
39	1.079	0.277	0.000	0.332	0.000	0.387	0.000	0.161
40	0.713	0.404	0.013	0.396	0.007	0.388	0.000	0.173

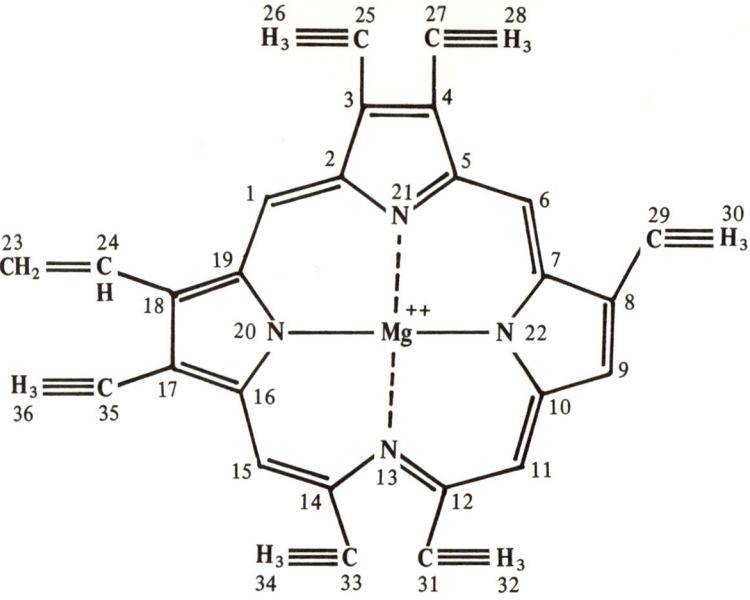

TABLE 31(P).—Chlorophylline.

Atom	P_{rr}	SDN_r	FOD_r	SDR_r	FRD_r	SDE_r	FED_r	π_{rr}
1	0.944	3.003	0.186	1.970	0.099	0.937	0.013	0.518
2	0.935	1.370	0.041	1.125	0.101	0.881	0.160	0.389
3	0.988	1.552	0.057	1.273	0.066	0.994	0.074	0.437
4	0.981	1.484	0.040	1.227	0.050	0.971	0.059	0.436
5	0.944	1.436	0.058	1.169	0.144	0.902	0.170	0.389
6	0.925	2.889	0.151	1.898	0.077	0.908	0.004	0.519
7	0.950	1.499	0.077	1.226	0.135	0.959	0.193	0.393
8	0.946	2.168	0.118	1.502	0.077	0.837	0.037	0.432
9	1.071	1.471	0.078	1.308	0.093	1.144	0.108	0.449
10	0.874	2.046	0.100	1.379	0.105	0.714	0.109	0.382
11	1.086	1.524	0.085	1.373	0.076	1.221	0.007	0.460
12	0.776	2.196	0.084	1.531	0.138	0.865	0.193	0.461
13	1.702	1.091	0.086	0.968	0.043	0.845	0.000	0.170
14	0.778	2.331	0.106	1.605	0.146	0.879	0.185	0.466
15	1.083	1.537	0.085	1.379	0.072	1.221	0.059	0.462
16	0.881	2.184	0.125	1.458	0.116	0.732	0.108	0.388
17	1.014	1.692	0.097	1.448	0.095	1.203	0.093	0.466
18	0.990	2.040	0.127	1.421	0.075	0.803	0.023	0.381
19	0.935	1.445	0.059	1.201	0.130	0.957	0.200	0.403
20	1.760	0.315	0.028	0.597	0.004	0.879	0.006	0.129
21	1.754	0.625	0.045	0.739	0.022	0.852	0.000	0.134
22	1.761	0.279	0.000	0.571	0.003	0.862	0.005	0.126
23	0.972	2.495	0.129	1.877	0.082	1.258	0.034	0.651
24	0.995	0.890	0.001	0.890	0.004	0.890	0.006	0.408
25	0.952	0.487	0.000	0.487	0.000	0.487	0.000	0.239
26	0.146	0.583	0.005	0.602	0.009	0.620	0.008	0.265
27	0.952	0.487	0.000	0.487	0.000	0.487	0.000	0.239
28	1.046	0.583	0.005	0.602	0.009	0.620	0.008	0.265

TABLE 31(P).—*Continued.*

29	0.952	0.487	0.000	0.487	0.000	0.487	0.000	0.239
30	1.041	0.666	0.014	0.635	0.010	0.603	0.005	0.266
31	0.955	0.486	0.000	0.486	0.000	0.486	0.000	0.239
32	1.028	0.680	0.010	0.645	0.019	0.605	0.027	0.268
33	0.955	0.486	0.000	0.486	0.000	0.486	0.000	0.239
34	1.028	0.680	0.010	0.645	0.019	0.605	0.027	0.268
35	0.952	0.487	0.000	0.487	0.000	0.487	0.000	0.239
36	1.050	0.608	0.012	0.628	0.012	0.648	0.013	0.266

TABLE 31(S).—Chlorophylline.

Atom	P_π	SDN_r	FOD_r	SDR_r	FRD_r	SDE_r	FED_r	π_π
1	0.946	2.818	0.192	1.881	0.103	0.944	0.014	0.517
2	0.940	1.299	0.038	1.096	0.101	0.893	0.163	0.388
3	0.988	1.496	0.062	1.253	0.069	1.011	0.077	0.439
4	0.980	1.419	0.041	1.201	0.049	0.894	0.058	0.438
5	0.949	1.372	0.060	1.145	0.117	0.917	0.175	0.389
6	0.925	2.691	0.149	1.800	0.076	0.910	0.003	0.517
7	0.964	1.431	0.081	1.203	0.138	0.975	0.195	0.393
8	0.929	2.044	0.117	1.442	0.078	0.843	0.038	0.435
9	1.079	1.357	0.077	1.257	0.092	1.157	0.107	0.444
10	0.873	1.917	0.098	1.317	0.104	0.716	0.110	0.381
11	1.096	1.435	0.089	1.337	0.078	1.239	0.067	0.456
12	0.766	2.070	0.079	1.470	0.136	0.870	0.194	0.461
13	1.702	0.986	0.085	0.914	0.042	0.842	0.000	0.168
14	0.769	2.234	0.110	1.562	0.151	0.890	0.192	0.468
15	1.091	1.459	0.089	1.350	0.077	1.241	0.065	0.459
16	0.885	2.084	0.132	1.412	0.121	0.740	0.110	0.389
17	1.009	1.632	0.101	1.429	0.101	1.225	0.101	0.473
18	0.995	1.883	0.126	1.346	0.074	0.809	0.021	0.379
19	0.937	1.377	0.059	1.173	0.131	0.970	0.205	0.403
20	1.761	0.307	0.003	0.594	0.005	0.881	0.006	0.128
21	1.756	0.569	0.044	0.712	0.022	0.855	0.000	0.132
22	1.763	0.262	0.000	0.566	0.002	0.864	0.005	0.125
23	0.974	2.341	0.129	1.804	0.080	1.267	0.031	0.649
24	0.994	0.891	0.001	0.891	0.003	0.891	0.005	0.408
25	1.079	0.287	0.000	0.332	0.000	0.387	0.000	0.161
26	0.914	0.416	0.004	0.398	0.004	0.379	0.006	0.173
27	1.079	0.278	0.000	0.332	0.000	0.387	0.000	0.161
28	0.914	0.416	0.004	0.398	0.004	0.379	0.006	0.173
29	1.079	0.278	0.000	0.332	0.000	0.387	0.000	0.161
30	0.910	0.454	0.008	0.411	0.006	0.369	0.003	0.173
31	1.080	0.277	0.000	0.331	0.001	0.385	0.001	0.160
32	0.901	0.454	0.006	0.412	0.010	0.370	0.015	0.173
33	1.080	0.277	0.000	0.331	0.001	0.385	0.001	0.160
34	0.901	0.454	0.006	0.412	0.010	0.370	0.015	0.173
35	1.079	0.277	0.000	0.333	0.000	0.388	0.001	0.161
36	0.915	0.425	0.007	0.411	0.007	0.396	0.008	0.173

TABLE 32(P).—Corrin.

Atom	P_{rr}	SDN_r	FOD_r	SDR_r	FRD_r	SDE_r	FED_r	π_{rr}
1	1.692	0.628	0.122	1.222	0.128	1.816	0.135	0.303
2	0.789	1.811	0.252	1.310	0.172	0.809	0.091	0.450
3	1.108	0.643	0.028	1.124	0.097	1.605	0.167	0.395
4	0.820	1.711	0.292	1.196	0.156	0.681	0.020	0.409
5	1.277	0.670	0.002	1.155	0.109	1.641	0.216	0.405
6	0.878	1.694	0.326	1.253	0.185	0.811	0.043	0.445
7	1.093	0.750	0.015	1.236	0.142	1.721	0.269	0.441
8	0.839	1.610	0.287	1.168	0.146	0.727	0.005	0.413
9	1.278	0.753	0.059	1.272	0.179	1.791	0.300	0.420
10	0.914	1.343	0.135	1.100	0.079	0.858	0.023	0.443
11	1.088	0.895	0.123	1.414	0.233	1.933	0.344	0.458
12	0.843	1.205	0.083	0.897	0.044	0.589	0.006	0.361
13	1.375	1.087	0.173	1.663	0.241	2.239	0.309	0.584
14	0.952	0.487	0.000	0.487	0.000	0.487	0.000	0.239
15	1.062	0.482	0.004	0.591	0.013	0.700	0.022	0.263
16	0.954	0.488	0.000	0.488	0.000	0.488	0.000	0.240
17	1.034	0.550	0.010	0.562	0.006	0.575	0.001	0.264
18	0.952	0.486	0.000	0.486	0.001	0.486	0.001	0.239
19	1.059	0.511	0.015	0.624	0.030	0.738	0.045	0.266
20	0.593	0.487	0.000	0.487	0.000	0.487	0.000	0.239
21	1.033	0.597	0.036	0.593	0.018	0.589	0.001	0.266
22	1.030	0.609	0.036	0.596	0.019	0.583	0.003	0.266
23	0.954	0.487	0.000	0.487	0.000	0.487	0.000	0.239

TABLE 32(S).—Corrin.

Atom	P_{rr}	SDN_r	FOD_r	SDR_r	FRD_r	SDE_r	FED_r	π_{rr}
1	1.697	0.596	0.125	1.221	0.131	1.847	0.137	0.299
2	0.796	1.758	0.266	1.296	0.180	0.834	0.094	0.453
3	1.108	0.629	0.025	1.136	0.097	1.643	0.168	0.397
4	0.815	1.659	0.299	1.172	0.159	0.685	0.020	0.409
5	1.283	0.649	0.002	1.162	0.110	1.675	0.219	0.400
6	0.874	1.643	0.337	1.231	0.190	0.819	0.044	0.445
7	1.123	0.732	0.018	1.245	0.144	1.758	0.271	0.437
8	0.836	1.558	0.290	1.146	0.148	0.734	0.006	0.414
9	1.258	0.721	0.058	1.280	0.181	1.839	0.304	0.415
10	0.920	1.297	0.135	1.086	0.080	0.876	0.025	0.443
11	1.087	0.877	0.127	1.436	0.239	1.994	0.350	0.462
12	0.838	1.191	0.088	0.892	0.047	0.593	0.007	0.362
13	1.384	1.022	0.172	1.653	0.242	2.284	0.312	0.570
14	1.080	0.277	0.000	0.333	0.000	0.390	0.001	0.161
15	0.922	0.355	0.002	0.391	0.007	0.427	0.013	0.173
16	1.080	0.276	0.000	0.331	0.000	0.386	0.000	0.161
17	0.905	0.394	0.006	0.373	0.003	0.351	0.001	0.172
18	1.079	0.277	0.000	0.333	0.000	0.390	0.001	0.161
19	0.921	0.372	0.009	0.412	0.017	0.451	0.026	0.173
20	1.080	0.276	0.000	0.331	0.000	0.386	0.000	0.161
21	0.904	0.419	0.020	0.390	0.010	0.360	0.000	0.173
22	1.080	0.276	0.000	0.331	0.000	0.386	0.000	0.161
23	0.903	0.426	0.021	0.391	0.011	0.357	0.001	0.173

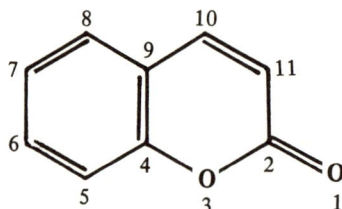

TABLE 33.—Coumarin.

Atom	P_{rr}	SDN$_r$	FOD$_r$	SDR$_r$	FRD$_r$	SDE$_r$	FED$_r$	π_{rr}
1	1.392	0.500	0.124	0.647	0.175	0.792	0.225	0.242
2	0.810	0.779	0.097	0.532	0.056	0.285	0.016	0.206
3	1.831	0.170	0.052	0.557	0.059	0.944	0.065	0.094
4	0.922	0.934	0.115	0.872	0.185	0.810	0.256	0.383
5	1.044	0.800	0.087	0.879	0.044	0.957	0.002	0.409
6	0.981	0.974	0.250	0.912	0.240	0.850	0.224	0.411
7	1.028	0.708	0.000	0.846	0.066	0.924	0.133	0.398
8	0.984	1.020	0.261	0.958	0.157	0.896	0.053	0.434
9	1.035	0.674	0.079	0.752	0.173	0.831	0.268	0.335
10	0.912	1.239	0.478	0.991	0.315	0.744	0.153	0.435
11	1.061	1.055	0.542	1.134	0.528	1.212	0.605	0.496

TABLE 34.—Cyanoriboflavin.

Atom	P_π	SDN_r	FOD_r	SDR_r	FRD_r	SDE_r	FED_r	π_π
1	1.011	0.856	0.013	0.903	0.082	0.951	0.152	0.414
2	0.980	1.103	0.160	0.942	0.094	0.780	0.028	0.384
3	1.028	1.002	0.101	1.018	0.097	1.033	0.093	0.443
4	0.918	1.023	0.093	0.862	0.099	0.700	0.106	0.358
5	1.563	0.775	0.187	1.079	0.253	1.382	0.319	0.328
6	0.830	1.168	0.108	0.825	0.054	0.482	0.001	0.324
7	1.459	0.516	0.066	1.229	0.421	1.942	0.775	0.359
8	0.826	0.721	0.021	0.508	0.019	0.296	0.016	0.205
9	1.745	0.164	0.003	0.832	0.006	1.500	0.009	0.151
10	0.814	0.831	0.053	0.554	0.029	0.277	0.006	0.205
11	0.967	1.364	0.295	1.112	0.221	0.861	0.147	0.412
12	1.032	2.008	0.532	1.362	0.271	0.716	0.010	0.455
13	0.996	0.708	0.001	0.755	0.070	0.803	0.139	0.337
14	0.983	1.220	0.190	1.058	0.096	0.897	0.001	0.443
15	0.848	0.758	0.010	0.569	0.005	0.381	0.000	0.253
16	1.172	0.683	0.038	0.651	0.022	0.620	0.005	0.287
17	1.442	0.427	0.037	0.670	0.089	0.914	0.141	0.229
18	1.379	0.631	0.093	0.694	0.071	0.758	0.050	0.250

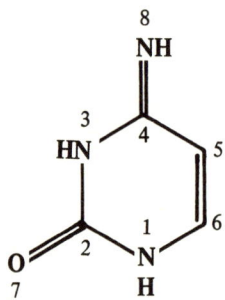

TABLE 35.—Cytosine (Imine form).

Atom	P_π	SDN_r	FOD_r	SDR_r	FRD_r	SDE_r	FED_r	π_π
1	1.659	0.315	0.184	0.948	0.266	1.581	0.347	0.232
2	0.809	0.700	0.004	0.489	0.004	0.279	0.003	0.204
3	1.701	0.348	0.128	0.881	0.086	1.514	0.045	0.197
4	0.843	0.966	0.417	0.723	0.211	0.480	0.004	0.311
5	1.178	0.578	0.120	1.090	0.320	1.603	0.519	0.406
6	0.839	1.947	0.850	1.004	0.507	0.761	0.164	0.428
7	1.436	0.340	0.004	0.589	0.015	0.837	0.026	0.217
8	1.536	0.473	0.292	1.456	0.591	2.439	0.890	0.408

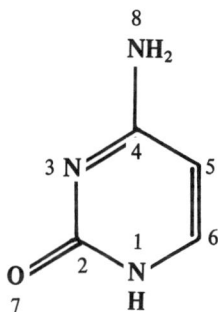

TABLE 36.—Cytosine (Lactam form 1).

Atom	P_π	SDN_r	FOD_r	SDR_r	FRD_r	SDE_r	FED_r	π_π
1	1.632	0.365	0.240	0.889	0.292	1.412	0.343	0.250
2	0.838	0.663	0.005	0.389	0.018	0.305	0.030	0.208
3	1.429	0.480	0.264	1.085	0.372	1.690	0.480	0.361
4	0.828	1.011	0.469	0.739	0.235	0.467	0.001	0.312
5	1.166	0.576	0.079	1.000	0.348	1.424	0.619	0.398
6	0.835	1.242	0.821	0.970	0.505	0.698	0.189	0.419
7	1.470	0.314	0.005	0.665	0.168	1.015	0.331	0.216
8	1.802	0.198	0.118	0.978	0.062	1.757	0.006	0.164

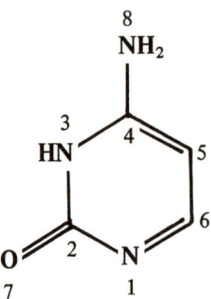

TABLE 37.—Cytosine (Lactam form 2).

Atom	P_π	SDN_r	FOD_r	SDR_r	FRD_r	SDE_r	FED_r	π_π
1	1.387	0.535	0.318	1.140	0.466	1.745	0.615	0.384
2	0.835	0.671	0.027	0.497	0.028	0.323	0.029	0.208
3	1.659	0.315	0.202	0.839	0.128	1.362	0.055	0.226
4	0.801	1.082	0.553	0.810	0.322	0.538	0.092	0.330
5	1.206	0.513	0.017	1.118	0.356	1.723	0.695	0.391
6	0.861	1.125	0.721	0.853	0.366	0.581	0.010	0.378
7	1.464	0.328	0.026	0.678	0.125	1.028	0.224	0.218
8	1.787	0.216	0.136	0.996	0.208	1.775	0.280	0.177

TABLE 38(P).—Deazaflavin.

Atom	P_π	SDN_r	FOD_r	SDR_r	FRD_r	SDE_r	FED_r	π_π
1	0.999	0.773	0.018	0.876	0.092	0.979	0.166	0.389
2	0.940	1.026	0.201	0.916	0.120	0.807	0.040	0.396
3	1.061	0.843	0.101	0.967	0.088	1.090	0.074	0.430
4	0.921	0.941	0.093	0.832	0.103	0.722	0.114	0.358
5	1.587	0.543	0.163	0.981	0.221	1.419	0.279	0.296
6	0.832	1.075	0.120	0.780	0.061	0.485	0.002	0.323
7	1.475	0.417	0.052	1.209	0.401	2.001	0.749	0.336
8	0.827	0.701	0.023	0.500	0.020	0.299	0.017	0.205
9	1.755	0.157	0.002	0.829	0.005	1.501	0.008	0.149
10	0.816	0.775	0.050	0.528	0.028	0.280	0.007	0.205
11	1.069	0.840	0.191	0.943	0.180	1.046	0.169	0.385
12	0.853	1.658	0.588	1.163	0.297	0.669	0.006	0.441
13	1.035	0.645	0.008	0.748	0.073	0.852	0.135	0.332
14	0.990	1.114	0.253	1.027	0.129	0.940	0.005	0.451
15	0.953	0.488	0.000	0.488	0.000	0.488	0.000	0.239
16	1.050	0.498	0.002	0.561	0.013	0.624	0.024	0.263
17	0.953	0.488	0.000	0.488	0.000	0.488	0.000	0.239
18	1.043	0.528	0.026	0.565	0.016	0.601	0.006	0.264
19	1.447	0.381	0.034	0.656	0.087	0.931	0.139	0.225
20	1.396	0.499	0.073	0.645	0.064	0.791	0.054	0.242

TABLE 38(S).—Deazaflavin.

Atom	P_{rr}	SDN_r	FOD_r	SDR_r	FRD_r	SDE_r	FED_r	π_{rr}
1	0.996	0.771	0.022	0.880	0.095	0.989	0.168	0.391
2	0.937	1.021	0.202	0.917	0.122	0.813	0.041	0.398
3	1.067	0.822	0.094	0.959	0.084	1.095	0.075	0.427
4	0.923	0.934	0.096	0.831	0.105	0.727	0.115	0.358
5	1.588	0.539	0.166	0.980	0.223	1.421	0.281	0.295
6	0.832	1.072	0.122	0.779	0.062	0.486	0.002	0.323
7	1.475	0.414	0.052	1.210	0.402	2.006	0.752	0.335
8	0.827	0.701	0.023	0.500	0.020	0.299	0.017	0.205
9	1.755	0.157	0.002	0.829	0.005	1.501	0.008	0.149
10	0.816	0.774	0.051	0.527	0.029	0.280	0.007	0.205
11	1.070	0.831	0.190	0.940	0.180	1.049	0.171	0.385
12	0.853	1.649	0.598	1.159	0.320	0.669	0.006	0.441
13	1.037	0.638	0.007	0.747	0.073	0.856	0.139	0.332
14	0.996	1.091	0.252	1.019	0.129	0.947	0.005	0.448
15	1.080	0.276	0.000	0.332	0.001	0.388	0.001	0.161
16	0.915	0.365	0.002	0.373	0.007	0.380	0.013	0.172
17	1.080	0.276	0.000	0.332	0.001	0.388	0.001	0.161
18	0.911	0.382	0.014	0.375	0.009	0.367	0.003	0.172
19	1.448	0.380	0.034	0.656	0.087	0.932	0.140	0.225
20	1.396	0.497	0.074	0.645	0.064	0.792	0.055	0.242

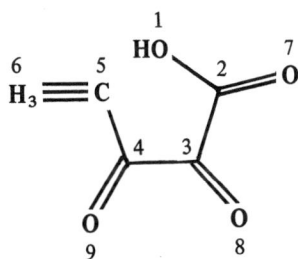

TABLE 39(P).—Dehydroascorbic Acid.

Atom	P_{rr}	SDN_r	FOD_r	SDR_r	FRD_r	SDE_r	FED_r	π_{rr}
1	1.919	0.069	0.019	0.505	0.681	0.941	1.371	0.045
2	0.775	0.882	0.159	0.568	0.106	0.254	0.052	0.203
3	0.792	1.056	0.450	0.661	0.226	0.226	0.003	0.211
4	0.783	0.985	0.271	0.625	0.136	0.266	0.001	0.214
5	0.954	0.941	0.004	0.491	0.008	0.491	0.012	0.242
6	1.027	0.531	0.038	0.537	0.026	0.543	0.015	0.259
7	1.306	0.626	0.192	0.586	0.365	0.547	0.538	0.242
8	1.208	0.984	0.542	0.720	0.285	0.456	0.029	0.260
9	1.235	0.805	0.326	0.640	0.167	0.475	0.008	0.251

TABLE 39(S).—Dehydroascorbic Acid.

Atom	P_{rr}	SDN_r	FOD_r	SDR_r	FRD_r	SDE_r	FED_r	π_{rr}
1	1.919	0.069	0.020	0.505	0.695	0.941	1.371	0.045
2	0.775	0.882	0.170	0.568	0.111	0.254	0.053	0.203
3	0.792	1.055	0.466	0.660	0.234	0.266	0.003	0.211
4	0.780	0.974	0.260	0.619	0.130	0.265	0.000	0.213
5	1.081	0.276	0.000	0.331	0.000	0.386	0.000	0.161
6	0.901	0.381	0.019	0.356	0.009	0.331	0.000	0.170
7	1.306	0.625	0.202	0.586	0.374	0.547	0.546	0.242
8	1.207	0.980	0.555	0.718	0.291	0.465	0.026	0.260
9	1.239	0.777	0.309	0.625	0.155	0.473	0.001	0.248

TABLE 40(P).—Dehydrobiopterin.

Atom	P_π	SDN_r	FOD_r	SDR_r	FRD_r	SDE_r	FED_r	π_π
1	1.676	0.263	0.045	0.822	0.024	1.391	0.003	0.207
2	0.778	1.080	0.252	0.834	0.154	0.589	0.057	0.322
3	1.399	0.409	0.005	1.084	0.070	1.759	0.136	0.332
4	0.864	0.990	0.333	0.883	0.210	0.775	0.087	0.348
5	1.221	0.455	0.005	1.558	0.292	2.662	0.578	0.360
6	0.842	0.690	0.010	0.551	0.023	0.411	0.036	0.215
7	1.447	0.360	0.011	0.806	0.086	1.251	0.161	0.230
8	1.172	0.858	0.388	0.964	0.238	1.071	0.087	0.405
9	0.996	1.057	0.592	1.692	0.552	2.327	0.512	0.583
10	0.954	0.485	0.011	0.485	0.006	0.485	0.001	0.238
11	1.051	0.525	0.088	0.653	0.077	0.781	0.066	0.270
12	0.954	0.485	0.011	0.485	0.006	0.485	0.001	0.238
13	1.051	0.525	0.088	0.653	0.077	0.781	0.066	0.270
14	1.787	0.208	0.069	1.009	0.076	1.810	0.082	0.172
15	1.807	0.198	0.092	1.112	0.108	2.025	0.125	0.165

TABLE 40(S).—Dehydrobiopterin.

Atom	P_π	SDN_r	FOD_r	SDR_r	FRD_r	SDE_r	FED_r	π_π
1	1.676	0.262	0.059	0.822	0.031	1.382	0.003	0.217
2	0.781	1.068	0.301	0.841	0.180	0.613	0.060	0.324
3	1.401	0.406	0.002	1.099	0.070	1.793	0.138	0.331
4	0.871	0.965	0.365	0.893	0.229	0.822	0.093	0.351
5	1.228	0.449	0.020	1.637	0.305	2.824	0.591	0.360
6	0.844	0.685	0.005	0.556	0.022	0.407	0.038	0.216
7	1.451	0.353	0.006	0.827	0.086	1.301	0.166	0.229
8	1.196	0.773	0.325	0.964	0.213	1.155	0.102	0.396
9	0.967	1.089	0.649	1.754	0.581	2.420	0.541	0.599
10	1.079	0.275	0.000	0.332	0.001	0.389	0.002	0.160
11	0.913	0.385	0.048	0.432	0.043	0.479	0.038	0.175
12	1.079	0.275	0.000	0.332	0.001	0.389	0.002	0.160
13	0.913	0.385	0.048	0.432	0.043	0.479	0.038	0.175
14	1.789	0.202	0.078	1.018	0.081	1.843	0.084	0.170
15	1.811	0.187	0.095	1.129	0.112	2.071	0.130	0.160

TABLE 41(P).—Dehydroluciferin.

Atom	P_{π}	SDN_r	FOD_r	SDR_r	FRD_r	SDE_r	FED_r	π_{π}
1	1.929	0.056	0.009	0.598	0.007	1.140	0.005	0.042
2	0.984	0.749	0.079	0.954	0.050	1.160	0.022	0.379
3	1.072	0.731	0.057	0.997	0.068	1.263	0.080	0.413
4	1.036	0.858	0.199	1.064	0.121	1.269	0.042	0.440
5	1.052	0.598	0.005	0.864	0.031	1.130	0.057	0.337
6	1.373	0.544	0.192	1.367	0.148	2.189	0.104	0.394
7	1.016	0.826	0.300	1.032	1.195	1.237	0.090	0.403
8	1.421	0.618	0.261	3.127	0.686	5.637	1.111	0.637
9	1.065	0.675	0.002	0.880	0.003	1.086	0.005	0.378
10	1.086	0.754	0.121	1.158	0.133	1.564	0.144	0.435
11	1.066	0.699	0.134	1.052	0.097	1.405	0.061	0.397
12	1.283	0.801	0.308	1.992	0.238	3.183	0.167	0.662
13	1.116	0.822	0.075	1.175	0.043	1.528	0.011	0.510
14	1.106	0.583	0.021	1.012	0.046	1.440	0.071	0.357
15	1.304	0.650	0.219	1.070	0.115	1.490	0.011	0.405
16	0.795	0.787	0.009	0.538	0.006	0.290	0.002	0.205
17	1.370	0.472	0.010	0.616	0.012	0.759	0.014	0.238
18	1.700	0.464	0.000	1.326	0.009	2.188	0.018	0.343

TABLE 41(F).—Dehydroluciferin.

Atom	P_{π}	SDN_r	FOD_r	SDR_r	FRD_r	SDE_r	FED_r	π_{π}
1	1.841	0.222	0.043	1.848	0.470	3.454	0.897	0.195
2	0.964	0.908	0.075	0.871	0.090	0.834	0.105	0.367
3	1.052	0.781	0.010	0.955	0.061	1.128	0.111	0.418
4	0.997	1.032	0.094	0.995	0.068	0.958	0.042	0.435
5	1.051	0.641	0.000	0.814	0.083	0.988	0.165	0.338
6	1.269	1.267	0.320	1.188	0.195	1.109	0.069	0.441
7	0.930	1.419	0.272	1.134	0.195	0.849	0.118	0.427
8	1.799	0.400	0.101	1.140	0.064	1.880	0.028	0.221
9	1.022	0.939	0.067	0.902	0.047	0.865	0.027	0.400
10	1.072	0.749	0.002	1.027	0.101	1.305	0.200	0.428
11	0.936	1.352	0.252	1.055	0.130	0.759	0.008	0.413
12	1.709	0.697	0.185	1.152	0.107	1.606	0.030	0.315
13	0.925	1.970	0.306	1.514	0.191	1.058	0.076	0.615
14	1.053	0.692	0.000	0.765	0.013	0.837	0.025	0.336
15	1.255	1.265	0.271	1.174	0.175	1.083	0.079	0.441
16	0.720	1.550	0.001	0.960	0.001	0.369	0.002	0.290
17	1.706	0.405	0.000	0.620	0.001	0.824	0.002	0.162
18	1.925	0.055	0.001	0.504	0.001	0.954	0.001	0.041

TABLE 42(P).—Diiodotyrosine (Alanyl side chain treated by −CH₃ group orbital).

Atom	P_{π}	SDN_r	FOD_r	SDR_r	FRD_r	SDE_r	FED_r	π_{π}
1	1.923	0.063	0.000	0.563	0.113	1.063	0.226	0.046
2	0.932	0.819	0.000	0.819	0.227	0.819	0.453	0.367
3	1.089	0.761	0.476	0.862	0.343	0.963	0.210	0.400
4	0.995	0.862	0.516	0.862	0.310	0.861	0.105	0.408
5	1.000	0.743	0.000	0.844	0.240	0.945	0.480	0.382
6	0.995	0.862	0.516	0.862	0.310	0.861	0.105	0.408
7	1.089	0.761	0.476	0.862	0.343	0.963	0.210	0.400
8	1.987	0.012	0.008	0.786	0.033	1.561	0.057	0.010
9	1.987	0.012	0.008	0.786	0.033	1.561	0.057	0.010
10	0.952	0.488	0.000	0.488	0.006	0.488	0.011	0.240
11	1.050	0.459	0.000	0.557	0.043	0.620	0.085	0.263

TABLE 42(S).—Diiodotyrosine (Alanyl side chain treated by −CH₃ group orbital).

Atom	P_{π}	SDN_r	FOD_r	SDR_r	FRD_r	SDE_r	FED_r	π_{π}
1	1.923	0.062	0.000	0.563	0.116	1.065	0.232	0.045
2	0.936	0.808	0.000	0.817	0.232	0.825	0.463	0.367
3	1.089	0.761	0.476	0.863	0.346	0.964	0.216	0.400
4	1.000	0.848	0.516	0.857	0.311	0.866	0.105	0.406
5	0.994	0.748	0.000	0.850	0.246	0.951	0.492	0.385
6	1.000	0.848	0.516	0.857	0.311	0.866	0.105	0.406
7	1.089	0.761	0.476	0.863	0.346	0.964	0.216	0.400
8	1.897	0.012	0.008	0.786	0.034	1.561	0.060	0.010
9	1.897	0.012	0.008	0.786	0.034	1.561	0.060	0.010
10	1.080	0.276	0.000	0.332	0.004	0.388	0.007	0.161
11	0.915	0.363	0.000	0.370	0.022	0.378	0.042	0.172

TABLE 43(I).—Dimethylbenzanthracene.

Atom	P_{rr}	SDN_r	FOD_r	SDR_r	FRD_r	SDE_r	FED_r	π_{rr}
1	1.016	0.838	0.032	0.950	0.059	1.063	0.086	0.411
2	1.015	0.851	0.133	0.964	0.115	1.076	0.096	0.413
3	0.998	0.976	0.157	1.089	0.148	1.201	0.139	0.449
4	1.049	0.667	0.016	0.780	0.031	0.892	0.046	0.337
5	0.841	0.934	0.137	1.384	0.262	1.834	0.387	0.474
6	1.074	0.668	0.009	0.868	0.038	1.068	0.067	0.354
7	0.996	0.782	0.074	0.832	0.069	0.882	0.065	0.352
8	1.014	0.899	0.035	0.049	0.023	0.999	0.012	0.430
9	1.005	0.878	0.146	0.928	0.099	0.978	0.051	0.411
10	1.009	0.836	0.002	0.886	0.145	0.936	0.028	0.405
11	1.009	0.946	0.128	0.996	0.081	1.046	0.033	0.440
12	1.006	0.721	0.065	0.771	0.056	0.821	0.046	0.336
13	1.033	0.917	0.227	1.117	0.186	1.317	0.146	0.452
14	0.989	1.011	0.291	1.061	0.195	1.111	0.099	0.447
15	1.067	0.633	0.024	0.833	0.057	1.033	0.090	0.339
16	0.832	0.999	0.291	1.448	0.357	1.899	0.423	0.492
17	1.051	0.651	0.011	0.763	0.024	0.876	0.037	0.335
18	0.996	0.991	0.223	1.103	0.186	1.216	0.149	0.451

TABLE 44(P).—DOPA.

Atom	P_π	SDN_r	FOD_r	SDR_r	FRD_r	SDE_r	FED_r	π_π
1	0.995	0.742	0.278	0.875	0.680	1.008	0.482	0.381
2	1.049	0.760	0.571	0.866	0.312	0.973	0.053	0.406
3	1.047	0.756	0.068	0.889	0.184	1.022	0.031	0.408
4	0.996	0.746	0.234	0.853	0.284	0.960	0.334	0.383
5	1.065	0.751	0.634	0.880	0.320	1.009	0.005	0.411
6	0.986	0.750	0.120	0.857	0.254	0.964	0.388	0.377
7	1.930	0.056	0.024	0.583	0.117	1.110	0.209	0.041
8	1.929	0.056	0.011	0.578	0.090	1.100	0.169	0.042
9	0.953	0.488	0.013	0.188	0.009	0.488	0.005	0.240
10	1.050	0.495	0.047	0.558	0.051	0.621	0.054	0.263

TABLE 44(S).—DOPA.

Atom	P_π	SDN_r	FOD_r	SDR_r	FRD_r	SDE_r	FED_r	π_π
1	0.999	0.731	0.145	0.974	0.316	1.017	0.488	0.380
2	1.049	0.760	0.644	0.867	0.348	0.975	0.053	0.407
3	1.053	0.742	0.215	0.885	0.260	1.028	0.305	0.406
4	0.989	0.751	0.099	0.859	0.218	0.966	0.338	0.386
5	1.070	0.738	0.602	0.875	0.304	1.012	0.005	0.409
6	0.986	0.751	0.253	0.858	0.324	0.965	0.394	0.377
7	1.930	0.054	0.013	0.583	0.112	1.112	0.212	0.041
8	1.929	0.056	0.022	0.578	0.097	1.100	0.172	0.042
9	1.080	0.276	0.000	0.332	0.002	0.388	0.004	0.161
10	0.915	0.363	0.008	0.371	0.018	0.379	0.028	0.172

TABLE 45.—Esculin.

Atom	P_{π}	SDN_r	FOD_r	SDR_r	FRD_r	SDE_r	FED_r	π_{rr}
1	1.007	0.712	0.007	0.886	0.152	1.060	0.298	0.378
2	0.957	0.893	0.236	0.921	0.262	0.949	0.288	0.391
3	1.096	0.709	0.050	0.901	0.026	1.094	0.001	0.410
4	0.945	0.882	0.123	0.910	0.196	0.938	0.270	0.390
5	1.834	0.164	0.054	0.562	0.055	0.963	0.056	0.092
6	0.812	0.769	0.103	0.532	0.061	0.295	0.018	0.207
7	1.083	0.977	0.440	1.151	0.413	1.326	0.386	0.474
8	0.913	1.213	0.522	0.976	0.280	0.739	0.038	0.423
9	1.060	0.614	0.051	0.789	0.160	0.963	0.269	0.335
10	1.041	0.917	0.258	0.987	0.131	1.058	0.003	0.446
11	1.922	0.079	0.028	0.584	0.068	1.090	0.107	0.048
12	1.932	0.052	0.001	0.587	0.056	1.123	0.111	0.040
13	1.397	0.481	0.127	0.656	0.142	0.831	0.156	0.241

TABLE 46.—5-Fluorouracil.

Atom	P_{π}	SDN_r	FOD_r	SDR_r	FRD_r	SDE_r	FED_r	π_{π}
1	1.676	0.264	0.161	0.911	0.358	1.557	0.555	0.212
2	0.808	0.701	0.023	0.489	0.017	0.278	0.010	0.203
3	1.750	0.167	0.075	0.826	0.039	1.485	0.003	0.153
4	0.827	0.715	0.189	0.503	0.102	0.292	0.016	0.206
5	1.255	0.571	0.331	1.094	0.598	1.618	0.866	0.405
6	0.838	1.170	1.022	0.935	0.606	0.700	0.189	0.413
7	1.981	0.016	0.011	0.378	0.042	0.739	0.074	0.009
8	1.436	0.341	0.020	0.587	0.068	0.833	0.115	0.217
9	1.431	0.353	0.167	0.639	0.170	0.885	0.174	0.231

Table 47(P).—Folic Acid.

Atom	P_π	SDN_r	FOD_r	SDR_r	FRD_r	SDE_r	FED_r	π_π
1	0.838	0.966	0.072	0.755	0.040	0.543	0.008	0.320
2	1.277	0.705	0.124	0.910	0.062	1.114	0.000	0.386
3	0.810	1.294	0.333	0.947	0.170	0.601	0.008	0.364
4	1.011	0.749	0.070	0.792	0.042	0.836	0.013	0.348
5	1.115	1.253	0.463	1.043	0.237	0.832	0.010	0.432
6	0.923	0.949	0.110	0.891	0.063	0.834	0.016	0.391
7	0.909	1.199	0.288	0.954	0.146	0.708	0.004	0.411
8	1.161	1.067	0.334	1.009	0.177	0.952	0.009	0.427
9	0.916	0.832	0.001	0.717	0.003	0.603	0.005	0.327
10	1.330	0.628	0.110	1.034	0.070	1.441	0.030	0.390
11	1.813	0.183	0.027	1.012	0.027	1.841	0.028	0.153
12	1.899	0.134	0.044	0.564	0.023	0.994	0.003	0.066
13	0.941	0.468	0.001	0.468	0.008	0.468	0.014	0.224
14	1.121	0.468	0.013	0.704	0.116	0.939	0.219	0.281
15	1.776	0.159	0.000	1.111	0.360	2.062	0.719	0.152
16	0.933	0.835	0.000	0.775	0.058	0.715	0.116	0.345
17	1.069	0.746	0.000	0.938	0.110	1.131	0.220	0.419
18	0.971	0.922	0.000	0.862	0.005	0.802	0.010	0.408
19	1.075	0.674	0.000	0.867	0.136	1.060	0.270	0.363
20	0.971	0.922	0.000	0.862	0.005	0.802	0.010	0.408
21	1.069	0.746	0.000	0.938	0.110	1.131	0.220	0.418
22	0.815	0.761	0.000	0.520	0.002	0.279	0.004	0.205
23	1.359	0.458	0.000	0.621	0.017	0.784	0.033	0.240
24	1.796	0.128	0.000	0.932	0.011	1.737	0.023	0.127
25	0.946	0.465	0.000	0.465	0.000	0.465	0.000	0.280
26	1.087	0.512	0.000	0.654	0.003	0.797	0.005	0.288
27	0.775	0.818	0.000	0.535	0.000	0.253	0.000	0.202
28	1.335	0.485	0.000	0.533	0.000	0.581	0.000	0.233
29	1.921	0.057	0.000	0.500	0.000	0.942	0.000	0.043

TABLE 47(S).—Folic Acid.

Atom	P_π	SDN_r	FOD_r	SDR_r	FRD_r	SDE_r	FED_r	π_π
1	0.839	0.962	0.070	0.754	0.040	0.545	0.010	0.329
2	1.277	0.705	0.126	0.910	0.063	1.114	0.000	0.386
3	0.811	1.288	0.333	0.945	0.171	0.603	0.010	0.364
4	1.012	0.748	0.071	0.793	0.044	0.839	0.017	0.348
5	1.121	1.227	0.453	1.032	0.232	0.838	0.012	0.430
6	0.916	0.950	0.106	0.893	0.063	0.836	0.020	0.392
7	0.913	1.195	0.303	0.954	0.153	0.713	0.004	0.412
8	1.160	1.067	0.347	1.009	0.179	0.952	0.012	0.427
9	0.918	0.820	0.001	0.716	0.004	0.606	0.006	0.327
10	1.330	0.629	0.114	1.036	0.076	1.443	0.039	0.391
11	1.814	0.181	0.026	1.113	0.031	1.844	0.035	0.152
12	1.899	0.134	0.044	0.564	0.024	0.995	0.004	0.065
13	1.070	0.272	0.000	0.330	0.006	0.388	0.002	0.156
14	0.949	0.365	0.008	0.450	0.054	0.535	0.100	0.184
15	1.804	0.150	0.000	1.145	0.375	2.139	0.750	0.143
16	0.931	0.831	0.000	0.771	0.064	0.712	0.127	0.343
17	1.072	0.741	0.000	0.942	0.116	1.143	0.232	0.418
18	0.971	0.992	0.000	0.861	0.006	0.801	0.011	0.408
19	1.077	0.668	0.000	0.870	0.144	1.071	0.289	0.363
20	0.971	0.922	0.000	0.861	0.006	0.801	0.011	0.408
21	1.072	0.741	0.000	0.942	0.116	1.143	0.232	0.418
22	0.816	0.755	0.000	0.517	0.002	0.278	0.004	0.205
23	1.399	0.448	0.000	0.619	0.018	0.789	0.036	0.238
24	1.825	0.114	0.000	0.949	0.011	1.184	0.023	0.113
25	1.071	0.270	0.000	0.327	0.000	0.384	0.000	0.155
26	0.932	0.375	0.000	0.417	0.001	0.459	0.002	0.183
27	0.767	0.808	0.000	0.530	0.000	0.252	0.000	0.155
28	1.338	0.451	0.000	0.512	0.000	0.953	0.000	0.227
29	1.921	0.054	0.000	0.498	0.000	0.941	0.000	0.043

TABLE 48(P).—N-5-Formyltetrahydrofolate.

Atom	P_{rr}	SDN_r	FOD_r	SDR_r	FRD_r	SDE_r	FED_r	π_{rr}
1	1.290	0.596	0.293	1.045	0.252	1.495	0.211	0.382
2	0.881	0.841	0.013	0.823	0.066	0.805	0.118	0.337
3	1.313	0.579	0.359	0.928	0.180	1.277	0.001	0.376
4	0.905	0.887	0.462	0.869	0.291	0.851	0.121	0.357
5	1.065	0.637	0.051	1.089	0.192	1.540	0.333	0.377
6	0.973	0.926	0.692	1.023	0.423	1.120	0.153	0.434
7	1.834	0.131	0.003	1.116	0.098	2.102	0.192	0.129
8	1.999	0.001	0.001	0.501	0.001	1.001	0.000	0.001
9	1.704	0.158	0.004	1.184	0.201	2.209	0.399	0.161
10	0.792	0.791	0.006	0.556	0.011	0.322	0.016	0.221
11	1.352	0.432	0.005	0.614	0.041	0.796	0.007	0.233
12	1.764	0.167	0.081	1.153	0.159	2.138	0.236	0.158
13	0.939	0.476	0.000	0.479	0.001	0.479	0.003	0.233
14	1.122	0.410	0.001	0.668	0.044	0.925	0.078	0.264
15	0.938	0.478	0.005	0.478	0.003	0.478	0.001	0.232
16	1.129	0.499	0.024	0.658	0.038	0.908	0.052	0.264

TABLE 48(S).—N-5-Formyltetrahydrofolate.

Atom	P_{rr}	SDN_r	FOD_r	SDR_r	FRD_r	SDE_r	FED_r	π_{rr}
1	1.293	0.589	0.285	1.056	0.253	1.522	0.219	0.382
2	0.882	0.839	0.016	0.829	0.069	0.819	0.122	0.338
3	1.319	0.571	0.358	0.927	0.180	1.284	0.001	0.374
4	0.904	0.880	0.452	0.870	0.288	0.859	0.124	0.355
5	1.065	0.635	0.061	1.103	0.202	1.572	0.343	0.376
6	0.978	0.917	0.701	1.033	0.431	1.149	0.160	0.435
7	1.835	0.130	0.003	1.122	0.100	2.114	0.197	0.129
8	1.999	0.001	0.001	0.501	0.001	1.001	0.000	0.001
9	1.727	0.152	0.005	1.234	0.212	2.315	0.418	0.156
10	0.974	0.785	0.008	0.555	0.012	0.325	0.017	0.221
11	1.357	0.426	0.007	0.620	0.044	0.815	0.081	0.232
12	1.792	0.161	0.094	1.199	0.172	2.236	0.250	0.151
13	1.073	0.277	0.000	0.336	0.001	0.394	0.003	0.160
14	0.954	0.322	0.001	0.419	0.020	0.516	0.040	0.174
15	1.072	0.277	0.000	0.335	0.001	0.393	0.001	0.160
16	0.957	0.322	0.009	0.415	0.016	0.508	0.024	0.175

TABLE 49.—N-10-Formyltetrahydrofolate.

Atom	P_{rr}	SDN_r	FOD_r	SDR_r	FRD_r	SDE_r	FED_r	π_{rr}
1	1.411	0.396	0.016	1.977	0.070	3.588	0.124	0.331
2	0.823	0.965	0.673	1.605	0.387	2.244	0.101	0.352
3	1.678	0.254	0.222	0.858	0.113	1.461	0.005	0.205
4	0.865	0.657	0.068	1.016	0.063	1.375	0.059	0.226
5	1.217	0.506	0.307	5.699	0.435	10.982	0.564	0.417
6	0.967	0.756	0.400	2.395	0.296	4.034	0.193	0.381
7	1.475	0.327	0.053	2.158	0.117	3.988	0.180	0.227
8	1.808	0.164	0.128	1.682	0.109	3.200	0.091	0.152
9	1.852	0.117	0.076	2.445	0.125	4.772	0.174	0.119
10	1.904	0.072	0.058	5.279	0.284	10.485	0.510	0.082

TABLE 50.—N-10-Formyltetrahydrofolate (Lactim form).

Atom	P_{rr}	SDN_r	FOD_r	SDR_r	FRD_r	SDE_r	FED_r	π_{rr}
1	1.341	0.517	0.316	0.947	0.167	1.377	0.018	0.362
2	0.888	0.826	0.002	0.896	0.069	0.967	0.136	0.342
3	1.333	0.521	0.282	1.026	0.999	1.530	0.116	0.363
4	0.931	0.852	0.653	1.063	0.417	1.274	0.181	0.394
5	1.101	0.580	0.007	1.342	0.205	2.105	0.403	0.378
6	0.923	0.835	0.569	0.960	0.360	1.085	0.152	0.362
7	1.921	0.068	0.058	0.611	0.052	1.153	0.047	0.048
8	1.837	0.126	0.000	1.183	0.093	2.240	0.186	0.126
9	1.836	0.140	0.112	1.241	0.160	2.343	0.209	0.134
10	1.889	0.078	0.001	1.695	0.277	3.313	0.552	0.087

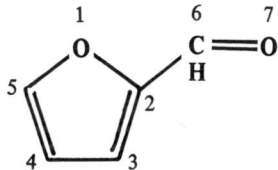

TABLE 51.—Fumaric Acid.

Atom	P_{rr}	SDN_r	FOD_r	SDR_r	FRD_r	SDE_r	FED_r	π_{rr}
1	1.923	0.071	0.020	0.509	0.011	0.946	0.001	0.043
2	0.791	0.872	0.155	0.565	0.078	0.258	0.001	0.201
3	0.944	1.449	0.610	1.141	0.714	0.834	0.818	0.479
4	0.944	1.449	0.610	1.141	0.714	0.834	0.818	0.479
5	0.791	0.872	0.155	0.565	0.078	0.258	0.001	0.201
6	1.923	0.071	0.020	0.509	0.011	0.946	0.001	0.043
7	1.341	0.672	0.215	0.652	0.197	0.632	0.179	0.149
8	1.341	0.672	0.215	0.652	0.197	0.632	0.179	0.249

TABLE 52.—2-Furaldehyde.

Atom	P_{rr}	SDN_r	FOD_r	SDR_r	FRD_r	SDE_r	FED_r	π_{rr}
1	1.752	0.276	0.183	0.537	0.092	0.798	0.000	0.134
2	1.048	0.786	0.380	0.998	0.539	1.209	0.698	0.426
3	1.046	0.796	0.333	0.873	0.278	0.949	0.223	0.405
4	1.102	0.610	0.028	0.822	0.152	1.033	0.276	0.375
5	0.970	1.002	0.499	1.078	0.584	1.155	0.668	0.491
6	0.793	0.890	0.285	0.589	0.147	0.287	0.009	0.222
7	1.288	0.617	0.291	0.613	0.208	0.608	0.125	0.250

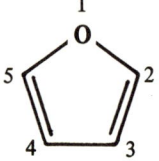

TABLE 53.—Furan.

Atom	P_{rr}	SDN_r	FOD_r	SDR_r	FRD_r	SDE_r	FED_r	π_{rr}
1	1.751	0.242	0.249	0.518	0.124	0.795	0.000	0.131
2	1.019	0.856	0.704	1.080	0.714	1.303	0.734	0.482
3	1.015	0.614	0.171	0.838	0.224	1.061	0.276	0.378
4	1.015	0.614	0.171	0.838	0.224	1.061	0.276	0.378
5	1.019	0.856	0.704	1.080	0.714	1.303	0.734	0.482

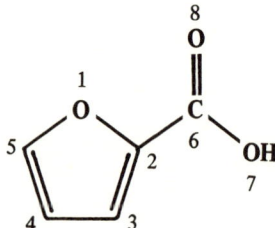

TABLE 54.—2-Furoic Acid.

Atom	P_{rr}	SDN_r	FOD_r	SDR_r	FRD_r	SDE_r	FED_r	π_{rr}
1	1.752	0.272	0.197	0.535	0.099	0.798	0.000	0.134
2	1.049	0.793	0.438	1.006	0.570	1.219	0.702	0.430
3	1.051	0.772	0.323	0.865	0.275	0.957	0.227	0.402
4	1.103	0.610	0.039	0.823	0.158	1.036	0.277	0.375
5	0.975	0.982	0.532	1.074	0.603	1.167	0.674	0.489
6	0.790	0.807	0.226	0.538	0.117	0.269	0.008	0.206
7	1.924	0.057	0.023	0.503	0.013	0.949	0.004	0.042
8	1.369	0.504	0.222	0.590	0.165	0.677	0.109	0.239

TABLE 55.—Guanine.

Atom	P_{rr}	SDN_r	FOD_r	SDR_r	FRD_r	SDE_r	FED_r	π_{rr}
1	0.808	0.988	0.698	0.838	0.404	0.688	0.110	0.335
2	1.681	0.237	0.133	0.842	0.085	1.438	0.044	0.202
3	0.835	0.701	0.000	0.530	0.015	0.359	0.029	0.212
4	1.171	0.503	0.106	1.115	0.232	1.727	0.358	0.354
5	1.288	0.550	0.032	0.921	0.087	1.293	0.142	0.360
6	0.988	0.855	0.344	1.168	0.331	1.481	0.318	0.462
7	1.593	0.341	0.235	0.799	0.143	1.258	0.051	0.245
8	0.975	0.735	0.269	0.874	0.211	1.013	0.154	0.360
9	1.419	0.391	0.048	1.272	0.254	2.153	0.461	0.332
10	1.803	0.170	0.134	1.049	0.160	1.927	0.186	0.155
11	1.428	0.370	0.000	0.728	0.074	1.085	0.147	0.231

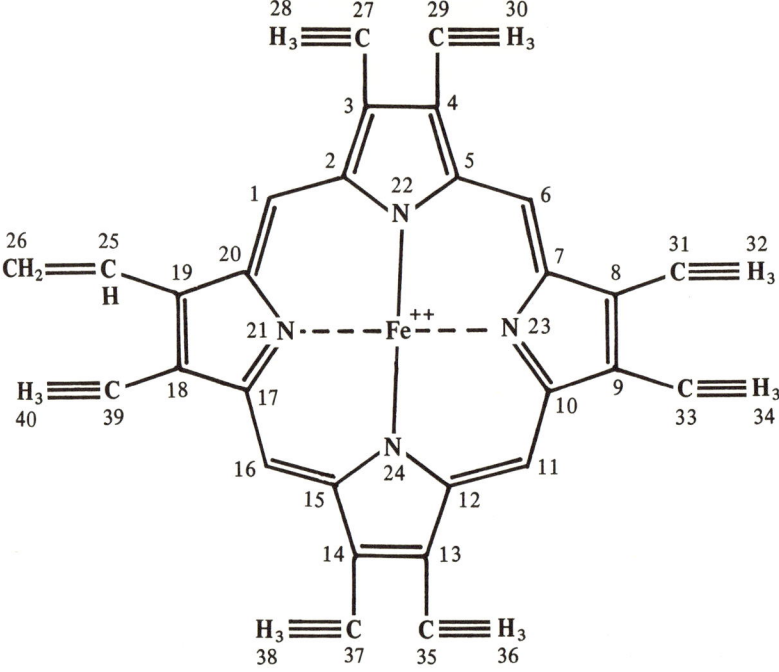

TABLE 56(P).—Heme.

Atom	P_{rr}	SDN_r	FOD_r	SDR_r	FRD_r	SDE_r	FED_r	π_{rr}
1	0.962	2.010	0.136	1.724	0.204	1.438	0.271	0.595
2	0.977	0.998	0.034	0.920	0.025	0.842	0.016	0.366
3	0.988	1.343	0.020	1.252	0.047	1.162	0.053	0.477
4	1.037	1.055	0.028	1.025	0.038	0.995	0.049	0.378
5	0.965	1.055	0.043	0.964	0.027	0.874	0.010	0.380
6	0.969	1.944	0.153	1.683	0.203	1.422	0.254	0.584
7	1.007	1.012	0.082	0.922	0.042	0.832	0.002	0.352
8	1.039	1.188	0.108	1.183	0.055	1.179	0.001	0.450
9	1.038	1.189	0.106	1.184	0.055	1.179	0.003	0.451
10	1.009	1.010	0.083	0.920	0.042	0.831	0.000	0.351
11	0.957	1.957	0.142	1.691	0.204	1.426	0.265	0.591
12	0.978	1.034	0.046	0.944	0.031	0.855	0.017	0.369
13	1.007	1.143	0.024	1.113	0.033	1.082	0.041	0.438
14	1.116	1.175	0.038	1.131	0.042	1.088	0.046	0.440
15	0.978	1.001	0.032	0.924	0.022	0.847	0.013	0.368
16	0.957	2.041	0.181	1.744	0.227	1.447	0.273	0.602
17	1.007	0.966	0.069	0.889	0.035	0.812	0.001	0.347
18	1.112	1.424	0.146	1.347	0.075	1.269	0.003	0.497
19	1.063	1.063	0.096	1.069	0.049	1.076	0.003	0.385
20	0.998	0.986	0.066	0.908	0.033	0.830	0.000	0.357
21	1.364	0.819	0.000	1.379	0.105	1.938	0.211	0.435
22	1.624	0.649	0.087	1.099	0.083	1.549	0.080	0.377
23	1.381	0.743	0.000	1.333	0.104	1.923	0.207	0.417
24	1.630	0.633	0.087	1.085	0.090	1.536	0.039	0.271
25	0.990	0.891	0.004	0.891	0.002	0.891	0.000	0.406
26	1.034	1.517	0.104	1.524	0.054	1.530	0.004	0.646
27	0.952	0.480	0.000	0.480	0.000	0.480	0.000	0.239
28	1.047	0.564	0.002	0.603	0.005	0.642	0.007	0.267

TABLE 56(P).—*Continued.*

29	0.993	0.894	0.001	0.894	0.003	0.894	0.005	0.407
30	1.015	1.517	0.030	1.488	0.044	1.458	0.057	0.641
31	0.952	0.487	0.000	0.487	0.000	0.487	0.000	0.239
32	1.053	0.547	0.013	0.596	0.007	0.646	0.000	0.265
33	0.952	0.487	0.000	0.487	0.000	0.487	0.000	0.239
34	1.053	0.547	0.013	0.596	0.007	0.646	0.000	0.265
35	0.952	0.487	0.000	0.487	0.000	0.487	0.000	0.239
36	1.050	0.541	0.003	0.588	0.004	0.634	0.005	0.265
37	0.952	0.487	0.000	0.487	0.000	0.487	0.000	0.239
38	1.049	0.545	0.005	0.590	0.005	0.635	0.006	0.265
39	0.952	0.486	0.000	0.486	0.000	0.486	0.000	0.239
40	1.050	0.574	0.018	0.614	0.009	0.655	0.000	0.267

TABLE 56(S).—Heme.

Atom	P_π	SDN_r	FOD_r	SDR_r	FRD_r	SDE_r	FED_r	π_π
1	0.962	1.957	0.134	1.700	0.203	1.443	0.271	0.593
2	0.981	0.983	0.034	0.915	0.024	0.847	0.015	0.365
3	0.979	1.340	0.019	1.256	0.067	1.172	0.054	0.483
4	1.044	1.015	0.030	1.012	0.040	1.009	0.050	0.376
5	0.966	1.046	0.045	0.962	0.028	0.877	0.010	0.381
6	0.971	1.883	0.155	1.659	0.206	1.435	0.257	0.582
7	1.009	1.004	0.086	0.920	0.044	0.836	0.002	0.352
8	1.042	1.143	0.106	1.175	0.054	1.206	0.002	0.452
9	1.040	1.145	0.104	1.175	0.054	1.205	0.003	0.453
10	1.012	1.001	0.089	0.918	0.044	0.836	0.000	0.351
11	0.957	1.898	0.143	1.666	0.205	1.434	0.266	0.589
12	0.980	1.023	0.048	0.941	0.032	0.858	0.016	0.369
13	1.008	1.112	0.025	1.108	0.034	1.104	0.044	0.441
14	1.006	1.149	0.042	1.128	0.045	1.107	0.048	0.443
15	0.972	0.986	0.031	0.919	0.022	0.853	0.013	0.367
16	0.956	1.995	0.188	1.723	0.230	1.451	0.273	0.600
17	1.010	0.950	0.071	0.884	0.036	0.817	0.001	0.346
18	1.003	1.417	0.151	1.349	0.077	1.281	0.003	0.505
19	1.071	1.003	0.090	1.048	0.047	1.092	0.003	0.381
20	0.999	0.971	0.068	0.903	0.034	0.835	0.001	0.357
21	1.366	0.807	0.000	1.378	0.105	1.950	0.209	0.435
22	1.626	0.625	0.087	1.089	0.083	1.554	0.079	0.275
23	1.386	0.718	0.000	1.332	0.104	1.946	0.207	0.413
24	1.633	0.606	0.087	1.073	0.089	1.539	0.091	0.267
25	0.990	0.892	0.004	0.892	0.002	0.892	0.000	0.406
26	1.040	1.460	0.099	1.504	0.051	1.548	0.004	0.463
27	1.079	0.277	0.000	0.332	0.000	0.387	0.000	0.161
28	0.913	0.404	0.001	0.398	0.003	0.392	0.004	0.174
29	0.993	0.895	0.001	0.895	0.003	0.895	0.004	0.408
30	1.020	1.479	0.032	1.476	0.046	1.473	0.059	0.638
31	1.079	0.277	0.000	0.332	0.000	0.388	0.000	0.161
32	0.917	0.391	0.007	0.393	0.005	0.395	0.000	0.173
33	1.079	0.277	0.000	0.332	0.000	0.388	0.000	0.161
34	0.917	0.391	0.007	0.393	0.005	0.395	0.000	0.173
35	1.079	0.277	0.000	0.332	0.000	0.388	0.000	0.161
36	0.915	0.388	0.002	0.388	0.003	0.388	0.003	0.173
37	1.079	0.277	0.000	0.332	0.000	0.388	0.000	0.161
38	0.915	0.391	0.003	0.390	0.003	0.388	0.004	0.173
39	1.079	0.277	0.000	0.332	0.000	0.388	0.000	0.161
40	0.915	0.409	0.011	0.405	0.005	0.400	0.000	0.174

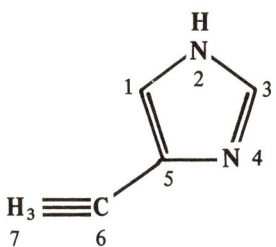

TABLE 57(P).—Histidine.

Atom	P_{rr}	SDN_r	FOD_r	SDR_r	FRD_r	SDE_r	FED_r	π_{rr}
1	1.090	0.700	0.501	1.036	0.610	1.365	0.719	0.444
2	1.548	0.389	0.447	0.738	0.243	1.087	0.040	0.261
3	0.991	0.837	0.801	0.992	0.691	1.047	0.581	0.447
4	1.269	0.515	0.187	0.797	0.136	1.080	0.083	0.342
5	1.065	0.597	0.048	0.880	0.269	1.163	0.491	0.371
6	0.952	0.488	0.004	0.488	0.006	0.488	0.007	0.240
7	1.058	0.487	0.011	0.562	0.045	0.647	0.079	0.262

TABLE 57(S).—Histidine.

Atom	P_{rr}	SDN_r	FOD_r	SDR_r	FRD_r	SDE_r	FED_r	π_{rr}
1	1.100	0.686	0.458	1.036	0.596	1.385	0.434	0.440
2	1.550	0.386	0.441	0.739	0.241	1.092	0.041	0.260
3	0.993	0.836	0.843	0.996	0.718	1.156	0.593	0.448
4	1.289	0.508	0.219	0.794	0.152	1.079	0.085	0.340
5	1.060	0.603	0.035	0.889	0.268	1.174	0.504	0.375
6	1.080	0.276	0.000	0.333	0.003	0.389	0.006	0.161
7	0.919	0.353	0.003	0.374	0.002	0.394	0.042	0.172

TABLE 58.—Hydroxystilbamidine.

Atom	P_{rr}	SDN_r	FOD_r	SDR_r	FRD_r	SDE_r	FED_r	π_{rr}
1	1.792	0.247	0.029	1.002	0.015	1.758	0.001	0.180
2	1.450	1.238	0.265	1.163	0.262	1.088	0.259	0.477
3	0.825	1.093	0.070	0.792	0.035	0.490	0.000	0.319
4	1.020	0.883	0.149	0.889	0.149	0.896	0.150	0.360
5	0.975	0.980	0.043	0.905	0.026	0.830	0.009	0.418
6	1.004	0.974	0.093	0.980	0.106	0.987	0.118	0.435
7	0.981	0.875	0.106	0.800	0.085	0.725	0.063	0.339
8	1.004	0.974	0.093	0.890	0.106	0.987	0.118	0.435
9	0.975	0.980	0.043	0.905	0.026	0.830	0.009	0.418
10	1.112	1.156	0.234	1.182	0.282	1.207	0.329	0.484
11	0.981	1.238	0.265	1.163	0.262	1.088	0.259	0.477
12	1.020	0.798	0.081	0.824	0.100	0.850	0.119	0.347
13	0.999	0.997	0.112	0.978	0.094	0.959	0.076	0.434
14	1.007	0.899	0.024	0.924	0.036	0.950	0.049	0.424
15	1.011	0.904	0.156	0.886	0.148	0.867	0.140	0.359
16	1.029	0.939	0.073	0.962	0.037	0.985	0.001	0.438
17	0.933	0.953	0.084	0.994	0.118	0.915	0.152	0.392
18	1.918	0.087	0.012	0.583	0.031	1.079	0.051	0.051
19	0.824	1.099	0.073	0.795	0.037	0.491	0.000	0.319
20	1.792	0.250	0.031	1.004	0.016	1.757	0.001	0.181
21	1.446	0.743	0.117	1.344	0.144	1.945	0.171	0.486

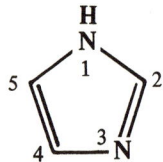

TABLE 59.—Imidazole.

Atom	P_{rr}	SDN_r	FOD_r	SDR_r	FRD_r	SDE_r	FED_r	π_{rr}
1	1.547	0.389	0.445	0.729	0.233	1.068	0.021	0.261
2	0.991	0.836	0.833	0.975	0.731	1.114	0.629	0.445
3	1.286	0.518	0.210	0.793	0.182	1.068	0.154	0.342
4	1.104	0.607	0.039	0.882	0.258	1.157	0.477	0.387
5	1.072	0.712	0.473	0.987	0.596	1.262	0.719	0.435

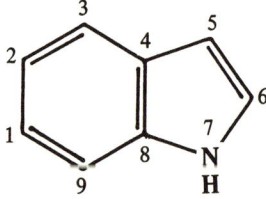

TABLE 60.—Indole.

Atom	P_{rr}	SDN_r	FOD_r	SDR_r	FRD_r	SDE_r	FED_r	π_{rr}
1	1.031	0.785	0.250	0.896	0.236	1.008	0.222	0.407
2	1.039	0.746	0.063	0.858	0.054	0.970	0.046	0.399
3	1.023	0.872	0.512	0.984	0.427	1.096	0.343	0.438
4	1.060	0.609	0.135	0.721	0.072	0.833	0.010	0.328
5	1.170	0.571	0.096	1.019	0.333	1.466	0.571	0.398
6	0.999	0.875	0.403	0.987	0.327	1.099	0.251	0.447
7	1.621	0.314	0.070	0.866	0.181	1.149	0.293	0.238
8	1.019	0.677	0.008	0.789	0.028	0.901	0.049	0.355
9	1.037	0.826	0.464	0.938	0.341	1.050	0.218	0.424

TABLE 61(P).—Iodotyrosine.

Atom	P_{rr}	SDN_r	FOD_r	SDR_r	FRD_r	SDE_r	FED_r	π_{rr}
1	1.049	0.770	0.433	0.872	0.317	0.974	0.202	0.408
2	0.942	0.804	0.010	0.817	0.235	0.829	0.460	0.367
3	1.089	0.759	0.533	0.861	0.381	0.963	0.228	0.400
4	1.001	0.845	0.451	0.858	0.274	0.871	0.097	0.407
5	0.999	0.744	0.005	0.845	0.251	0.947	0.497	0.383
6	1.006	0.841	0.556	0.853	0.339	0.866	0.122	0.405
7	1.924	0.062	0.001	0.564	0.116	1.067	0.231	0.045
8	1.987	0.012	0.009	0.786	0.036	1.561	0.063	0.010
9	0.952	0.488	0.000	0.488	0.006	0.488	0.012	0.240
10	1.050	0.495	0.001	0.557	0.045	0.620	0.088	0.263

TABLE 61(S).—Iodotyrosine.

Atom	P_{rr}	SDN_r	FOD_r	SDR_r	FRD_r	SDE_r	FED_r	π_{rr}
1	1.049	0.771	0.445	0.873	0.326	0.975	0.208	0.409
2	0.946	0.793	0.007	0.815	0.239	0.836	0.471	0.366
3	1.088	0.760	0.525	0.862	0.380	0.964	0.236	0.400
4	1.006	0.831	0.472	0.854	0.284	0.876	0.097	0.405
5	0.993	0.749	0.003	0.851	0.256	0.953	0.509	0.386
6	1.011	0.828	0.539	0.849	0.332	0.840	0.124	0.403
7	1.925	0.060	0.001	0.564	0.119	1.069	0.237	0.044
8	1.987	0.011	0.099	0.786	0.037	1.561	0.063	0.010
9	1.080	0.276	0.000	0.332	0.004	0.388	0.008	0.161
10	0.915	0.363	0.000	0.371	0.022	0.378	0.045	0.172

TABLE 62.—Isobergapten.

Atom	P_{rr}	SDN_r	FOD_r	SDR_r	FRD_r	SDE_r	FED_r	π_{rr}
1	1.113	0.650	0.014	0.951	0.152	1.247	0.290	0.402
2	0.996	0.915	0.001	1.140	0.230	1.365	0.461	0.493
3	1.790	0.199	0.024	0.653	0.042	0.928	0.061	0.115
4	0.970	0.831	0.179	0.872	0.123	0.912	0.067	0.381
5	1.124	0.658	0.001	0.983	0.141	1.307	0.281	0.413
6	0.938	0.915	0.162	0.956	0.209	0.997	0.255	0.399
7	1.091	0.565	0.029	0.783	0.015	1.001	0.001	0.332
8	0.907	1.195	0.583	0.962	0.292	0.930	0.001	0.419
9	1.098	0.916	0.449	1.134	0.225	1.352	0.000	0.487
10	0.813	0.760	0.106	0.527	0.054	0.294	0.002	0.207
11	1.826	0.178	0.073	0.572	0.070	0.966	0.066	0.098
12	0.918	1.014	0.224	1.004	0.227	0.995	0.231	0.418
13	1.095	0.558	0.011	0.783	0.097	1.008	0.183	0.329
14	1.919	0.080	0.019	0.588	0.053	1.096	0.088	0.050
15	1.402	0.463	0.124	0.650	0.069	0.836	0.013	0.239

TABLE 63.—Isopimpinellin.

Atom	P_{rr}	SDN_r	FOD_r	SDR_r	FRD_r	SDE_r	FED_r	π_{rr}
1	0.992	0.921	0.011	1.146	0.120	0.371	0.229	0.491
2	1.116	0.653	0.023	0.979	0.110	1.306	0.196	0.403
3	1.091	0.559	0.000	0.784	0.051	1.009	0.101	0.328
4	0.953	0.980	0.293	1.170	0.341	1.360	0.388	0.448
5	1.092	0.562	0.030	0.800	0.038	1.037	0.046	0.333
6	0.906	1.209	0.577	0.970	0.292	0.731	0.007	0.420
7	0.916	1.008	0.125	0.857	0.102	1.262	0.036	0.417
8	0.812	0.767	0.115	0.528	0.058	0.289	0.000	0.207
9	1.840	0.152	0.052	0.594	0.058	1.037	0.065	0.089
10	0.991	0.835	0.108	1.027	0.159	1.220	0.210	0.413
11	1.058	0.664	0.028	1.070	0.200	1.476	0.373	0.395
12	1.006	0.792	0.147	0.897	0.076	1.003	0.004	0.391
13	1.794	0.192	0.009	0.595	0.049	0.998	0.089	0.114
14	1.917	0.091	0.034	0.630	0.076	1.168	0.117	0.045
15	1.937	0.048	0.003	0.631	0.058	1.213	0.113	0.037

TABLE 64.—Khellin.

Atom	P_{rr}	SDN_r	FOD_r	SDR_r	FRD_r	SDE_r	FED_r	π_{rr}
1	0.991	0.924	0.094	1.149	0.169	1.374	0.245	0.491
2	1.116	0.648	0.046	0.947	0.116	1.301	0.186	0.402
3	1.090	0.504	0.019	0.789	0.067	1.014	0.116	0.329
4	0.954	0.896	0.295	1.086	0.313	1.276	0.332	0.422
5	1.116	0.592	0.018	0.829	0.020	1.067	0.022	0.354
6	0.837	0.770	0.243	0.531	0.122	0.292	0.001	0.207
7	1.153	0.669	0.140	1.027	0.117	1.386	0.095	0.430
8	0.838	1.233	0.620	0.994	0.313	0.755	0.006	0.444
9	1.804	0.197	0.074	0.640	0.085	1.083	0.097	0.114
10	0.990	0.779	0.003	0.972	0.122	1.165	0.241	0.394
11	1.056	0.677	0.112	1.083	0.238	1.489	0.365	0.398
12	1.009	0.750	0.048	0.856	0.024	0.961	0.001	0.383
13	1.795	0.187	0.001	0.590	0.041	0.994	0.080	0.113
14	1.920	0.077	0.031	0.615	0.065	1.154	0.100	0.050
15	1.936	0.050	0.012	0.632	0.061	1.215	0.109	0.038
16	1.397	0.509	0.246	0.678	0.126	0.847	0.006	0.254

TABLE 65.—Kynurenic Acid.

Atom	P_{rr}	SDN_r	FOD_r	SDR_r	FRD_r	SDE_r	FED_r	π_{rr}
1	1.007	0.875	0.125	0.889	0.146	0.902	0.168	0.407
2	0.986	0.915	0.093	0.872	0.088	0.828	0.083	0.404
3	1.013	0.992	0.226	1.005	0.304	1.019	0.341	0.444
4	0.965	0.749	0.001	0.705	0.003	0.662	0.004	0.329
5	1.192	1.006	0.385	1.059	0.389	1.112	0.394	0.448
6	0.944	1.036	0.304	0.862	0.171	0.688	0.039	0.364
7	1.053	0.816	0.001	0.941	0.143	1.067	0.285	0.429
8	0.885	1.117	0.324	0.943	0.298	0.769	0.271	0.388
9	1.117	0.720	0.008	0.734	0.008	0.747	0.007	0.339
10	0.985	1.029	0.243	0.985	0.265	0.942	0.287	0.438
11	1.909	0.110	0.041	0.975	0.077	1.040	0.113	0.058
12	0.785	0.849	0.085	0.552	0.043	0.256	0.001	0.201
13	1.922	0.945	0.505	0.065	0.000	0.011	0.006	0.043
14	1.337	0.592	0.113	0.601	0.060	0.611	0.007	0.243

TABLE 66(P).—Luciferin.

Atom	$P_{\pi\pi}$	SDN$_r$	FOD$_r$	SDR$_r$	FRD$_r$	SDE$_r$	FED$_r$	$\pi_{\pi\pi}$
1	1.837	0.236	0.950	1.813	0.531	3.389	1.012	0.203
2	0.959	0.931	0.090	0.865	0.093	0.799	0.096	0.366
3	1.051	0.785	0.012	0.948	0.062	1.111	0.113	0.418
4	0.991	1.061	0.113	0.995	0.073	0.928	0.033	0.435
5	1.051	0.639	0.000	0.802	0.082	0.965	0.164	0.337
6	1.237	1.385	0.381	1.183	0.208	0.980	0.035	0.438
7	0.936	1.478	0.334	1.181	0.233	0.884	0.133	0.441
8	1.791	0.435	0.121	1.150	0.088	1.866	0.055	0.232
9	1.016	0.963	0.079	0.896	0.047	0.830	0.015	0.398
10	1.072	0.748	0.003	1.025	0.120	1.302	0.238	0.428
11	1.787	0.462	0.135	1.719	0.075	2.977	0.014	0.322
12	0.867	1.468	0.293	0.998	0.147	0.526	0.001	0.358
13	1.400	1.204	0.350	1.455	0.215	1.705	0.080	0.522
14	1.069	0.266	0.000	0.321	0.000	0.375	0.001	0.152
15	0.895	0.554	0.037	0.514	0.023	0.473	0.009	0.195
16	0.667	1.544	0.000	0.946	0.000	0.348	0.000	0.287
17	1.861	0.384	0.000	0.585	0.000	0.780	0.000	0.170
18	1.695	0.370	0.000	1.223	0.000	2.076	0.001	0.313

TABLE 66(F).—Luciferin.

Atom	$P_{\pi\pi}$	SDN$_r$	FOD$_r$	SDR$_r$	FRD$_r$	SDE$_r$	FED$_r$	$\pi_{\pi\pi}$
1	1.835	0.247	0.051	1.815	0.528	3.382	1.000	0.207
2	0.958	0.943	0.087	0.870	0.091	0.797	0.095	0.366
3	1.051	0.788	0.013	0.949	0.062	1.110	0.112	0.418
4	0.990	1.073	0.109	1.000	0.071	0.927	0.033	0.435
5	1.050	0.641	0.000	0.802	0.082	0.962	0.163	0.337
6	1.235	1.440	0.375	1.210	0.205	0.980	0.035	0.411
7	0.937	1.482	0.306	1.182	0.219	0.882	0.132	0.439
8	1.791	0.447	0.117	1.157	0.086	1.868	0.055	0.233
9	1.015	0.975	0.078	0.902	0.047	0.829	0.015	0.399
10	1.072	0.748	0.002	1.024	0.119	1.301	0.236	0.428
11	1.775	0.525	0.148	1.735	0.081	2.945	0.014	0.347
12	0.864	1.557	0.307	1.046	0.154	0.535	0.001	0.365
13	1.396	1.153	0.305	1.400	0.193	1.648	0.080	0.488
14	0.953	0.451	0.002	0.451	0.002	0.451	0.002	0.213
15	1.022	0.900	0.094	0.857	0.058	0.814	0.021	0.332
16	0.685	1.590	0.003	0.971	0.002	0.352	0.000	0.289
17	1.685	0.412	0.001	0.603	0.001	0.784	0.000	0.172
18	1.685	0.424	0.002	1.248	0.001	2.073	0.001	0.336

TABLE 67(P).—Lumichrome.

Atom	P_{π}	SDN_r	FOD_r	SDR_r	FRD_r	SDE_r	FED_r	π_{rr}
1	0.978	0.863	0.067	0.891	0.152	0.919	0.238	0.394
2	0.953	0.986	0.144	0.896	0.073	0.825	0.002	0.393
3	1.040	0.967	0.162	1.031	0.217	1.095	0.273	0.450
4	0.957	0.818	0.443	0.247	0.055	0.675	0.067	0.335
5	1.226	1.016	0.299	1.101	0.337	1.186	0.376	0.446
6	0.873	0.973	0.059	0.799	0.099	0.626	0.139	0.342
7	1.721	0.206	0.009	0.918	0.214	1.631	0.418	0.179
8	0.805	0.711	0.009	0.492	0.007	0.272	0.006	0.203
9	1.752	0.167	0.009	0.829	0.008	1.491	0.006	0.151
10	0.811	0.804	0.058	0.537	0.030	0.270	0.001	0.204
11	0.982	1.091	0.280	0.962	0.157	0.832	0.035	0.393
12	1.089	1.456	0.502	1.141	0.287	0.828	0.091	0.451
13	0.992	0.707	0.003	0.735	0.031	0.764	0.059	0.335
14	1.021	1.077	0.227	1.042	0.241	1.007	0.175	0.453
15	0.953	0.488	0.000	0.488	0.000	0.488	0.000	0.239
16	1.045	0.522	0.018	0.563	0.009	0.604	0.000	0.264
17	0.953	0.488	0.000	0.488	0.000	0.488	0.000	0.239
18	1.048	0.509	0.009	0.562	0.023	0.616	0.037	0.263
19	1.430	0.361	0.013	0.584	0.036	0.808	0.058	0.219
20	1.379	0.553	0.088	0.645	0.052	0.737	0.015	0.246

TABLE 67(S).—Lumichrome.

Atom	P_π	SDN_r	FOD_r	SDR_r	FRD_r	SDE_r	FED_r	π_{rr}
1	0.975	0.862	0.071	0.894	0.156	0.926	0.240	0.396
2	0.950	0.966	0.148	0.898	0.075	0.830	0.002	0.395
3	1.040	0.940	0.154	1.022	0.215	1.012	0.275	0.446
4	0.958	0.814	0.046	0.746	0.057	0.678	0.068	0.335
5	1.227	1.008	0.302	1.098	0.341	1.188	0.379	0.336
6	0.874	0.968	0.058	0.798	0.099	0.625	0.140	0.342
7	1.721	0.205	0.009	0.919	0.215	1.633	0.422	0.179
8	0.805	0.711	0.009	0.492	0.007	0.272	0.006	0.203
9	1.752	0.166	0.009	0.829	0.008	1.491	0.006	0.151
10	0.811	0.803	0.059	0.537	0.030	0.270	0.001	0.204
11	0.983	1.081	0.280	0.958	0.158	0.835	0.035	0.393
12	1.090	1.446	0.510	1.137	0.301	0.829	0.092	0.451
13	0.993	0.703	0.004	0.735	0.032	0.767	0.059	0.335
14	1.019	1.051	0.222	1.032	0.200	1.014	0.177	0.450
15	1.080	0.276	0.000	0.332	0.000	0.387	0.000	0.161
16	0.912	0.387	0.010	0.373	0.005	0.369	0.000	0.172
17	1.080	0.276	0.000	0.332	0.000	0.387	0.003	0.161
18	0.913	0.371	0.005	0.373	0.012	0.376	0.020	0.172
19	1.431	0.360	0.013	0.584	0.036	0.808	0.058	0.219
20	1.380	0.551	0.089	0.644	0.052	0.738	0.015	0.246

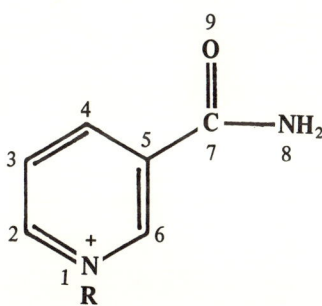

TABLE 68.—Nicotinamide Adenine Dinucleotide, oxidized (NAD$^+$).

Atom	P_π	SDN$_r$	FOD$_r$	SDR$_r$	FRD$_r$	SDE$_r$	FED$_r$	π_π
1	1.577	1.053	0.354	0.839	0.177	0.625	0.000	0.219
2	0.856	2.147	0.549	1.367	0.178	0.587	0.007	0.447
3	0.936	0.999	0.012	0.843	0.007	0.687	0.002	0.389
4	0.798	2.082	0.609	1.302	0.305	0.521	0.001	0.403
5	0.957	0.971	0.027	0.815	0.017	0.660	0.007	0.355
6	0.836	2.134	0.435	1.354	0.222	0.574	0.009	0.448
7	0.808	0.791	0.004	0.527	0.003	0.263	0.002	0.202
8	1.852	0.125	0.002	0.912	0.848	1.698	1.695	0.112
9	1.380	0.513	0.007	0.614	0.141	0.716	0.276	0.243

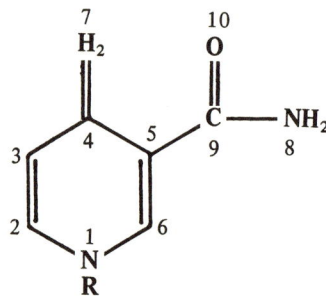

TABLE 69(P).—Nicotinamide Adenine Dinucleotide, reduced (NADH).

Atom	P_{rr}	SDN_r	FOD_r	SDR_r	FRD_r	SDE_r	FED_r	π_{rr}
1	1.643	0.295	0.043	1.253	0.261	2.211	0.479	0.242
2	0.975	0.689	0.185	0.977	0.150	1.084	1.115	0.430
3	1.127	0.750	0.314	1.526	0.418	2.302	0.523	0.526
4	0.953	0.469	0.000	0.469	0.003	0.469	0.006	0.228
5	1.168	0.602	0.314	1.379	0.390	2.155	0.466	0.428
6	0.893	1.099	0.742	0.964	0.391	0.829	0.041	0.435
7	1.122	0.438	0.000	0.869	0.130	1.299	0.261	0.281
8	1.864	0.099	0.043	0.934	0.033	1.769	0.024	0.098
9	0.828	0.724	0.186	0.520	0.100	0.316	0.015	0.207
10	1.426	0.412	0.174	0.679	0.122	0.947	0.071	0.235

TABLE 69(S).—Nicotinamide Adenine Dinucleotide, reduced (NADH).

Atom	P_{rr}	SDN_r	FOD_r	SDR_r	FRD_r	SDE_r	FED_r	π_{rr}
1	1.659	0.273	0.043	1.350	0.275	2.427	0.507	0.233
2	0.989	0.842	0.185	1.003	0.157	1.164	0.128	0.431
3	1.126	0.761	0.314	1.633	0.430	2.506	0.546	0.542
4	1.075	0.272	0.000	0.334	0.005	0.397	0.010	0.156
5	1.171	0.605	0.314	1.478	0.402	2.351	0.490	0.438
6	0.900	1.086	0.742	0.974	0.394	0.862	0.047	0.437
7	0.956	0.343	0.000	0.591	0.077	0.839	0.155	0.185
8	1.865	0.098	0.043	0.938	0.034	1.778	0.025	0.098
9	0.830	0.721	0.186	0.524	0.101	0.326	0.016	0.208
10	1.428	0.410	0.174	0.695	0.125	0.980	0.075	0.235

TABLE 70.—Orotic Acid.

Atom	P_{π}	SDN_r	FOD_r	SDR_r	FRD_r	SDE_r	FED_r	π_{π}
1	0.892	1.147	0.591	0.934	0.392	0.700	0.193	0.383
2	1.139	0.865	0.427	1.192	0.632	1.519	0.837	0.499
3	0.827	0.738	0.116	0.514	0.065	0.290	0.015	0.206
4	1.747	0.185	0.055	0.833	0.029	1.482	0.002	0.158
5	0.808	0.699	0.008	0.488	0.010	0.287	0.011	0.203
6	1.687	0.302	0.135	0.951	0.372	1.599	0.609	0.220
7	1.922	0.067	0.026	0.506	0.014	0.945	0.001	0.043
8	0.787	0.860	0.230	0.560	0.117	0.259	0.003	0.202
9	1.331	0.617	0.267	0.616	0.152	0.615	0.037	0.247
10	1.421	0.446	0.136	0.658	0.152	0.870	0.171	0.238
11	1.438	0.339	0.010	0.588	0.066	0.837	0.122	0.217

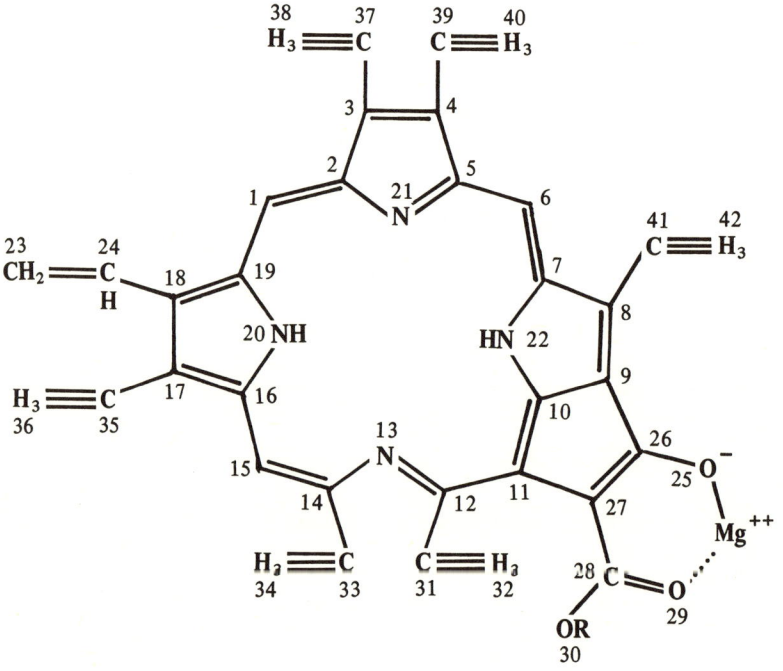

TABLE 71.—Peripheral Mg^{++}—complex of Pheophytin A.

Atom	P_{rr}	SDN_r	FOD_r	SDR_r	FRD_r	SDE_r	FED_r	π_{rr}
1	0.964	2.065	0.196	1.802	0.177	1.539	0.133	0.596
2	1.021	0.845	0.012	0.871	0.006	0.897	0.000	0.350
3	1.055	0.968	0.009	1.110	0.008	1.253	0.006	0.441
4	1.044	1.035	0.010	1.117	0.005	1.200	0.000	0.446
5	1.034	0.775	0.011	0.860	0.008	0.546	0.006	0.347
6	0.894	2.488	0.204	1.844	0.163	1.250	0.123	0.538
7	1.022	0.805	0.034	0.890	0.017	0.975	0.000	0.361
8	0.892	2.191	0.241	1.538	0.121	0.904	0.001	0.485
9	1.125	0.772	0.456	0.906	0.028	1.037	0.003	0.346
10	1.038	0.947	0.006	1.262	0.080	1.577	0.154	0.429
11	1.089	1.052	0.104	1.355	0.150	1.658	0.197	0.408
12	1.249	1.096	1.000	1.375	0.131	1.653	0.161	0.447
13	0.878	1.276	0.004	1.140	0.011	1.003	0.010	0.442
14	0.896	1.329	0.062	1.134	0.032	0.938	0.002	0.423
15	1.077	1.216	0.085	1.489	0.117	1.762	0.149	0.515
16	0.935	1.238	0.091	1.044	0.053	0.848	0.015	0.376
17	1.021	1.248	0.098	1.274	0.060	1.300	0.023	0.472
18	1.027	1.248	0.111	1.120	0.070	0.991	0.030	0.384
19	0.987	0.946	0.036	0.971	0.022	0.997	0.007	0.381
20	1.636	0.447	0.007	1.037	0.031	1.627	0.054	0.257
21	1.355	1.096	0.136	1.491	0.119	1.886	0.102	0.455
22	1.642	0.450	0.007	1.232	0.095	2.014	0.184	0.264
23	1.005	1.706	0.118	1.578	0.075	1.449	0.032	0.469
24	0.993	0.892	0.004	0.892	0.003	0.892	0.001	0.407
25	1.817	0.452	0.058	1.244	0.100	2.035	0.142	0.158
26	0.911	1.545	0.137	1.470	0.158	1.396	0.178	0.457
27	1.242	0.492	0.007	1.332	0.100	2.171	0.193	0.354
28	0.808	0.790	0.001	0.565	0.006	0.339	0.011	0.216
29	1.367	0.477	0.002	0.685	0.023	0.892	0.044	0.242
30	1.966	0.025	0.000	0.504	0.001	0.984	0.001	0.019
31	0.953	0.487	0.000	0.487	0.000	0.487	0.000	0.239
32	1.038	0.556	0.001	0.590	0.001	0.623	0.002	0.266
33	0.953	0.187	0.000	0.187	0.000	0.487	0.000	0.239
34	1.040	0.564	0.008	0.590	0.004	0.616	0.000	0.265
35	0.953	0.487	0.000	0.487	0.000	0.487	0.000	0.239
36	1.015	0.553	0.012	0.606	0.007	0.660	0.003	0.266
37	0.953	0.487	0.000	0.487	0.000	0.487	0.000	0.239
38	1.055	0.520	0.001	0.588	0.001	0.655	0.001	0.265
39	0.953	0.487	0.000	0.487	0.000	0.487	0.000	0.239
40	1.055	0.520	0.001	0.588	0.001	0.655	0.000	0.265
41	0.953	0.487	0.000	0.487	0.000	0.487	0.000	0.239
42	1.036	0.666	0.030	0.638	0.015	0.609	0.000	0.268

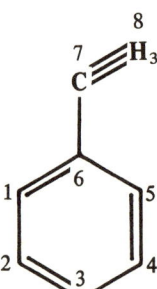

TABLE 72(P).—Phenylalanine (Toluene) (Alanyl side chain treated by $-CH_3$ group orbital).

Atom	P_{rr}	SDN_r	FOD_r	SDR_r	FRD_r	SDE_r	FED_r	π_{rr}
1	1.015	0.825	0.165	0.850	0.183	0.875	0.201	0.405
2	0.999	0.832	0.149	0.832	0.318	0.832	0.126	0.398
3	1.010	0.810	0.616	0.845	0.608	0.870	0.600	0.401
4	0.999	0.832	0.149	0.832	0.318	0.832	0.126	0.398
5	1.015	0.825	0.164	0.850	0.183	0.875	0.201	0.405
6	0.963	0.814	0.615	0.814	0.602	0.814	0.588	0.378
7	0.953	0.488	0.028	0.488	0.028	0.488	0.028	0.240
8	1.046	0.504	0.116	0.554	0.123	0.604	0.130	0.263

TABLE 72(S).—Phenylalanine (Toluene) (Alanyl side chain treated by $-CH_3$ group orbital).

Atom	P_{rr}	SDN_r	FOD_r	SDR_r	FRD_r	SDE_r	FED_r	π_{rr}
1	1.020	0.811	0.500	0.845	0.352	0.879	0.201	0.403
2	0.999	0.833	0.500	0.833	0.318	0.833	0.135	0.398
3	1.013	0.808	0.000	0.842	0.314	0.875	0.629	0.400
4	0.999	0.833	0.500	0.833	0.318	0.833	0.135	0.398
5	1.020	0.811	0.500	0.845	0.352	0.879	0.201	0.403
6	0.957	0.818	0.000	0.818	0.310	0.818	0.619	0.381
7	1.080	0.276	0.000	0.332	0.007	0.387	0.014	0.161
8	0.912	0.368	0.000	0.368	0.030	0.368	0.061	0.172

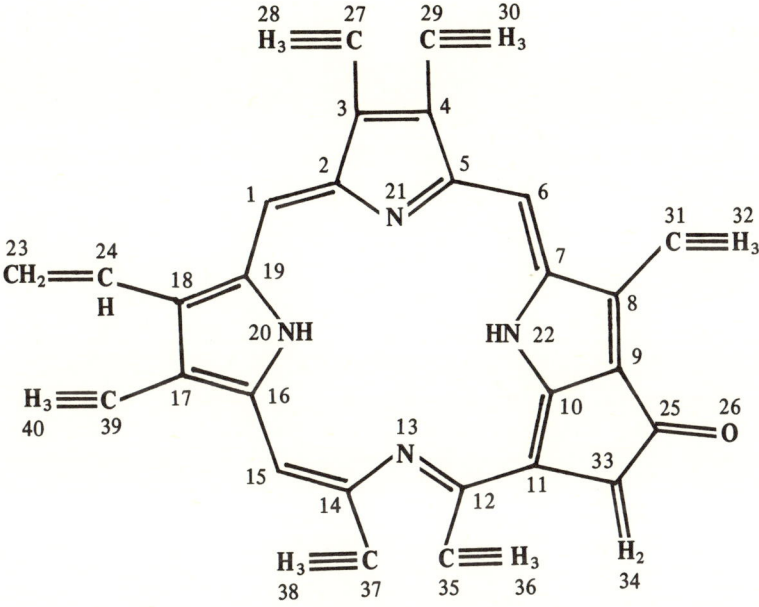

TABLE 73(P).—Pheophytin *A*.

Atom	P_{rr}	SDN_r	FOD_r	SDR_r	FRD_r	SDE_r	FED_r	π_{rr}
1	0.976	1.783	0.224	1.617	0.234	1.451	0.244	0.548
2	1.025	0.808	0.005	0.864	0.003	0.920	0.001	0.349
3	1.053	0.981	0.021	1.107	0.012	1.234	0.002	0.442
4	1.048	0.995	0.002	1.109	0.002	1.223	0.002	0.444
5	1.031	0.792	0.024	0.860	0.012	0.928	0.001	0.347
6	0.947	1.864	0.123	1.632	0.185	1.399	0.247	0.586
7	1.009	0.839	0.059	0.907	0.035	0.975	0.012	0.365
8	0.941	1.454	0.124	1.210	0.075	0.967	0.027	0.455
9	1.103	0.892	0.093	1.004	0.062	1.117	0.031	0.385
10	0.951	1.054	0.021	0.958	0.024	0.862	0.021	0.381
11	1.037	1.190	0.163	1.426	0.219	1.662	0.274	0.494
12	0.911	1.134	0.006	1.046	0.004	0.960	0.002	0.421
13	1.244	1.131	0.176	1.355	0.215	1.579	0.254	0.463
14	0.907	1.149	0.038	1.046	0.019	0.942	0.000	0.416
15	1.072	1.216	0.147	1.443	0.204	1.671	0.260	0.520
16	0.963	1.066	0.085	0.963	0.050	0.860	0.015	0.373
17	1.021	1.175	0.142	1.231	0.096	1.287	0.050	0.470
18	1.033	1.093	0.129	1.043	0.088	0.992	0.046	0.381
19	0.990	0.889	0.036	0.945	0.023	1.001	0.011	0.038
20	1.637	0.432	0.005	1.008	0.042	1.584	0.078	0.255
21	1.375	0.814	0.119	1.362	0.156	1.910	0.194	0.426
22	1.630	0.433	0.005	1.005	0.059	1.577	0.113	0.258
23	1.012	1.554	0.147	1.503	0.101	1.453	0.054	0.643
24	0.993	0.893	0.009	0.893	0.007	0.893	0.004	0.407
25	0.802	0.899	0.017	0.591	0.009	0.283	0.002	0.214
26	1.304	0.684	0.032	0.663	0.020	0.642	0.009	0.256
27	0.952	0.487	0.000	0.487	0.000	0.487	0.000	0.239
28	1.055	0.522	0.002	0.587	0.001	0.653	0.000	0.265

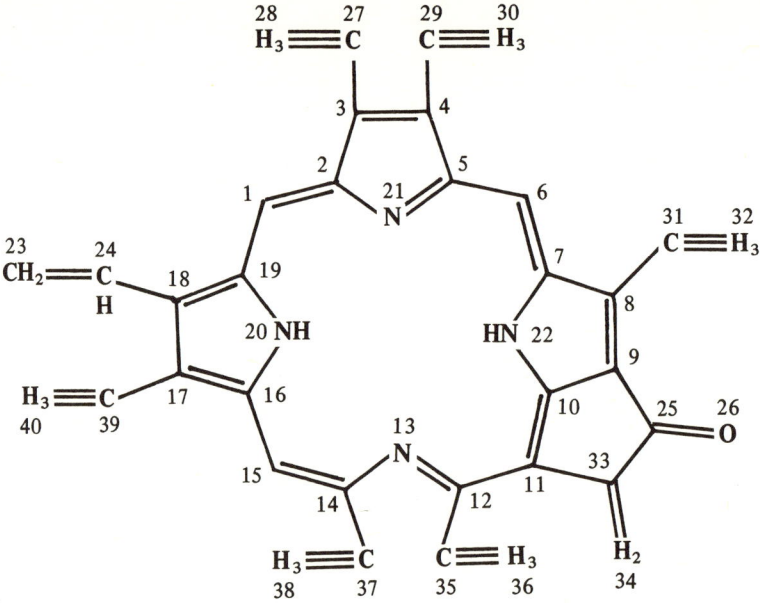

TABLE 73(P).—*Continued.*

29	0.952	0.487	0.000	0.487	0.000	0.487	0.000	0.239
30	1.055	0.522	0.002	0.587	0.001	0.653	0.000	0.265
31	0.952	0.487	0.000	0.487	0.000	0.487	0.000	0.239
32	1.042	0.578	0.015	0.598	0.009	0.619	0.004	0.266
33	0.958	0.480	0.000	0.480	0.000	0.480	0.001	0.233
34	1.030	0.577	0.009	0.623	0.020	0.669	0.030	0.275
35	0.953	0.487	0.000	0.487	0.000	0.487	0.000	0.239
36	1.041	0.540	0.001	0.580	0.001	0.619	0.000	0.265
37	0.953	0.487	0.000	0.487	0.000	0.487	0.000	0.239
38	1.041	0.542	0.005	0.580	0.002	0.617	0.000	0.265
39	0.952	0.486	0.000	0.486	0.000	0.486	0.000	0.239
40	1.051	0.544	0.170	0.601	0.012	0.658	0.007	0.266

TABLE 73(S).—Pheophytin *A.*

Atom	P_{π}	SDN_r	FOD_r	SDR_r	FRD_r	SDE_r	FED_r	π_{rr}
1	0.978	1.727	0.247	1.601	0.238	1.475	0.247	0.577
2	1.029	0.797	0.004	0.864	0.003	0.931	0.001	0.349
3	1.056	0.947	0.021	1.105	0.012	1.264	0.002	0.444
4	1.051	0.961	0.002	1.106	0.002	1.252	0.002	0.446
5	1.035	0.782	0.026	0.861	0.013	0.941	0.001	0.347
6	0.949	1.807	0.117	1.612	0.183	1.419	0.249	0.584
7	1.014	0.828	0.062	0.907	0.037	0.987	0.011	0.364
8	0.933	1.438	0.122	1.207	0.075	0.977	0.028	0.460
9	1.110	0.849	0.090	0.988	0.061	1.126	0.032	0.379
10	0.954	1.029	0.019	0.950	0.023	0.872	0.028	0.380
11	1.037	1.161	0.166	1.429	0.221	1.698	0.277	0.496
12	0.907	1.122	0.006	1.050	0.004	0.977	0.003	0.424

TABLE 73(S).—*Continued.*

13	1.249	0.090	0.175	1.340	0.214	1.590	0.253	0.456
14	0.901	1.140	0.040	1.049	0.020	0.958	0.000	0.419
15	1.079	1.183	0.152	1.438	0.206	1.694	0.259	0.516
16	0.966	1.053	0.091	0.962	0.053	0.871	0.015	0.373
17	1.015	1.174	0.147	1.241	0.099	1.308	0.051	0.477
18	1.040	1.048	0.127	1.028	0.088	1.007	0.048	0.378
19	0.992	0.879	0.036	0.946	0.023	1.013	0.011	0.038
20	1.640	0.420	0.006	1.006	0.042	1.592	0.078	0.253
21	1.381	0.783	0.116	1.360	0.155	1.938	0.194	0.422
22	1.634	0.418	0.006	1.005	0.060	1.592	0.114	0.255
23	1.017	1.511	0.147	1.490	0.101	1.470	0.056	0.641
24	0.993	0.894	0.010	0.894	0.007	0.894	0.004	0.407
25	0.800	0.887	0.018	0.585	0.010	0.283	0.002	0.213
26	1.310	0.660	0.033	0.654	0.022	0.649	0.010	0.253
27	1.079	0.277	0.000	0.332	0.000	0.388	0.000	0.161
28	0.918	0.377	0.003	0.388	0.001	0.399	0.000	0.173
29	1.079	0.277	0.000	0.332	0.000	0.388	0.000	0.161
30	0.918	0.377	0.003	0.388	0.001	0.399	0.000	0.173
31	1.079	0.277	0.000	0.332	0.000	0.388	0.000	0.161
32	0.910	0.411	0.009	0.394	0.005	0.378	0.002	0.173
33	1.079	0.272	0.000	0.327	0.001	0.383	0.001	0.157
34	0.899	0.431	0.005	0.413	0.011	0.410	0.017	0.177
35	1.080	0.276	0.000	0.331	0.000	0.387	0.000	0.161
36	0.909	0.389	0.000	0.384	0.000	0.378	0.000	0.173
37	1.080	0.276	0.000	0.332	0.000	0.387	0.000	0.161
38	0.909	0.390	0.003	0.384	0.001	0.377	0.000	0.173
39	1.079	0.277	0.000	0.332	0.000	0.388	0.000	0.161
40	0.915	0.392	0.010	0.397	0.007	0.402	0.004	0.174

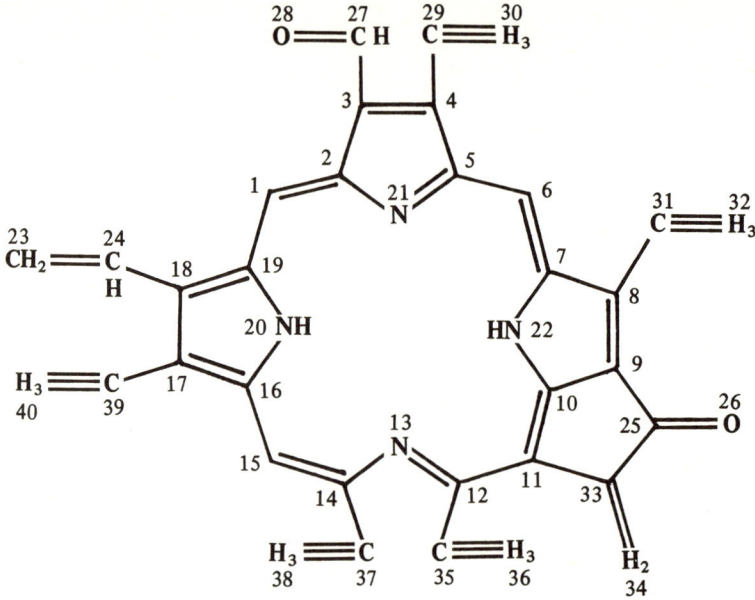

TABLE 74(P).—Pheophytin *B*.

Atom	P_{rr}	SDN_r	FOD_r	SDR_r	FRD_r	SDE_r	FED_r	π_{rr}
1	0.981	1.868	0.283	1.666	0.264	1.466	0.245	0.585
2	1.008	0.812	0.002	0.850	0.002	0.887	0.001	0.350
3	1.117	0.888	0.037	1.008	0.019	1.128	0.000	0.388
4	0.971	1.344	0.006	1.205	0.004	1.066	0.002	0.469
5	1.032	0.777	0.032	0.844	0.016	0.911	0.000	0.343
6	0.938	1.955	0.052	1.665	0.148	1.376	0.243	0.591
7	1.009	0.835	0.055	0.902	0.034	0.969	0.013	0.364
8	0.937	1.480	0.072	1.215	0.049	0.950	0.026	0.455
9	1.103	0.890	0.079	1.003	0.055	1.116	0.032	0.384
10	0.947	1.079	0.007	0.962	0.016	0.846	0.025	0.381
11	1.037	1.201	0.158	1.431	0.217	1.661	0.275	0.495
12	0.908	1.157	0.000	1.052	0.002	0.946	0.003	0.421
13	1.243	1.143	0.161	1.358	0.207	1.574	0.252	0.464
14	0.904	1.191	0.057	1.059	0.029	0.926	0.000	0.416
15	1.071	1.226	0.130	1.443	0.193	1.659	0.257	0.520
16	0.959	1.114	0.107	0.981	0.062	0.849	0.016	0.374
17	1.019	1.200	0.142	1.238	0.095	1.276	0.049	0.471
18	1.032	1.139	0.154	1.064	0.101	0.989	0.047	0.382
19	0.984	0.902	0.032	0.939	0.021	0.977	0.010	0.379
20	1.638	0.435	0.011	1.010	0.045	1.585	0.078	0.254
21	1.355	0.878	0.124	1.355	0.163	1.833	0.202	0.435
22	1.629	0.451	0.012	1.013	0.062	1.575	0.113	0.260
23	1.010	1.599	0.174	1.524	0.115	1.449	0.055	0.646
24	0.993	0.893	0.010	0.893	0.007	0.893	0.004	0.407
25	0.809	0.899	0.014	0.591	0.008	0.283	0.002	0.214
26	1.304	0.684	0.026	0.663	0.018	0.642	0.009	0.256
27	0.796	0.896	0.007	0.589	0.003	0.282	0.000	0.218
28	1.303	0.662	0.013	0.643	0.007	0.623	0.000	0.250
29	0.952	0.487	0.000	0.487	0.000	0.487	0.000	0.239
30	1.046	0.565	0.001	0.597	0.000	0.630	0.000	0.267

TABLE 74(P).—*Continued.*

31	0.952	0.487	0.000	0.487	0.000	0.487	0.000	0.239
32	1.042	0.581	0.009	0.599	0.006	0.616	0.003	0.266
33	0.958	0.480	0.000	0.480	0.000	0.480	0.000	0.233
34	1.030	0.579	0.010	0.624	0.020	0.668	0.030	0.275
35	0.953	0.487	0.000	0.487	0.000	0.487	0.000	0.239
36	1.041	0.543	0.000	0.580	0.000	0.617	0.000	0.265
37	0.953	0.487	0.000	0.487	0.000	0.487	0.000	0.239
38	1.041	0.547	0.007	0.581	0.004	0.615	0.000	0.265
39	0.952	0.486	0.000	0.486	0.000	0.486	0.000	0.239
40	1.051	0.547	0.017	0.602	0.012	0.657	0.006	0.266

TABLE 74(S).—Pheophytin *B.*

Atom	P_{rr}	SDN_r	FOD_r	SDR_r	FRD_r	SDE_r	FED_r	π_{rr}
1	0.985	1.806	0.286	1.649	0.267	1.491	0.248	0.583
2	1.009	0.804	0.002	0.850	0.002	0.896	0.001	0.351
3	1.125	0.842	0.038	0.987	0.020	1.137	0.001	0.383
4	0.963	1.329	0.008	1.203	0.005	1.078	0.002	0.475
5	1.037	0.765	0.034	0.844	0.017	0.922	0.000	0.343
6	0.938	1.896	0.057	1.644	0.146	1.391	0.245	0.589
7	1.014	0.823	0.057	0.901	0.035	0.980	0.012	0.365
8	0.928	1.463	0.070	1.211	0.048	0.959	0.026	0.459
9	1.110	0.848	0.078	0.996	0.054	1.124	0.033	0.379
10	0.950	1.053	0.005	0.954	0.016	0.854	0.026	0.380
11	1.037	1.172	0.158	1.434	0.218	1.700	0.278	0.496
12	0.904	1.144	0.000	1.054	0.002	0.964	0.003	0.424
13	1.248	1.103	0.161	1.344	0.206	1.584	0.252	0.457
14	0.897	1.182	0.059	1.061	0.030	0.940	0.000	0.419
15	1.077	1.197	0.135	1.438	0.195	1.679	0.257	0.517
16	0.963	1.102	0.113	0.981	0.065	0.860	0.017	0.374
17	1.012	1.203	0.147	1.249	0.098	1.295	0.049	0.478
18	1.039	1.093	0.151	1.049	0.100	1.005	0.050	0.380
19	0.986	0.895	0.033	0.941	0.021	0.987	0.009	0.380
20	1.641	0.421	0.013	1.007	0.045	1.592	0.077	0.252
21	1.358	0.856	0.124	1.354	0.163	1.853	0.203	0.433
22	1.633	0.436	0.014	1.013	0.064	1.590	0.113	0.257
23	1.015	1.554	0.172	1.510	0.115	1.466	0.058	0.643
24	0.993	0.894	0.011	0.894	0.008	0.894	0.004	0.407
25	0.800	0.889	0.015	0.585	0.008	0.283	0.002	0.213
26	1.310	0.659	0.028	0.654	0.019	0.648	0.010	0.253
27	0.796	0.890	0.007	0.586	0.004	0.282	0.000	0.218
28	1.304	0.648	0.014	0.637	0.007	0.625	0.000	0.250
29	1.079	0.276	0.000	0.332	0.000	0.387	0.000	0.161
30	0.912	0.403	0.001	0.394	0.000	0.385	0.000	0.173
31	1.079	0.277	0.000	0.332	0.000	0.387	0.000	0.161
32	0.910	0.413	0.005	0.395	0.003	0.377	0.002	0.173
33	1.079	0.272	0.000	0.327	0.001	0.383	0.001	0.157
34	0.899	0.414	0.005	0.412	0.011	0.410	0.017	0.177
35	1.080	0.276	0.000	0.331	0.000	0.387	0.000	0.161
36	0.909	0.390	0.000	0.384	0.000	0.377	0.000	0.173
37	1.080	0.276	0.000	0.331	0.000	0.387	0.000	0.161
38	0.909	0.393	0.004	0.384	0.002	0.376	0.000	0.173
39	1.079	0.277	0.000	0.332	0.000	0.388	0.000	0.161
40	0.915	0.394	0.010	0.398	0.007	0.401	0.004	0.174

TABLE 75(P).—Pheophytin D.

Atom	P_{rr}	SDN_r	FOD_r	SDR_r	FRD_r	SDE_r	FED_r	π_{rr}
1	0.974	1.899	0.256	1.695	0.271	1.490	0.285	0.597
2	1.021	0.815	0.002	0.854	0.001	0.892	0.000	0.348
3	1.050	1.014	0.031	1.120	0.018	1.227	0.005	0.445
4	1.046	1.006	0.000	1.107	0.000	1.207	0.001	0.445
5	1.026	0.822	0.031	0.866	0.018	0.911	0.004	0.348
6	0.948	1.858	0.097	1.621	0.175	1.384	0.254	0.584
7	1.006	0.866	0.063	0.910	0.035	0.955	0.006	0.365
8	0.941	1.460	0.109	1.214	0.072	0.968	0.035	0.456
9	1.101	0.917	0.094	1.008	0.060	1.099	0.025	0.385
10	0.950	1.054	0.017	0.958	0.025	0.862	0.033	0.381
11	1.028	1.260	0.174	1.426	0.215	1.589	0.255	0.497
12	0.911	1.133	0.004	1.045	0.002	0.956	0.001	0.421
13	1.235	1.204	0.184	1.358	0.214	1.512	0.244	0.466
14	0.908	1.163	0.044	1.055	0.023	0.946	0.001	0.417
15	1.060	1.289	0.155	1.443	0.206	1.597	0.257	0.523
16	0.967	1.088	0.093	0.979	0.053	0.871	0.013	0.375
17	0.971	1.341	0.170	1.206	0.099	1.071	0.029	0.457
18	1.072	1.128	0.140	1.077	0.087	1.025	0.034	0.398
19	0.891	0.871	0.027	0.910	0.026	0.948	0.025	0.370
20	1.629	0.462	0.011	1.004	0.063	1.546	0.116	0.261
21	1.376	0.829	0.116	1.374	0.166	1.919	0.216	0.427
22	1.628	0.445	0.008	1.000	0.058	1.554	0.108	0.260
23	0.793	0.931	0.026	0.605	0.014	0.281	0.002	0.218
24	1.292	0.738	0.049	0.671	0.029	0.603	0.008	0.251
25	0.802	0.902	0.016	0.192	0.009	0.282	0.002	0.213
26	1.303	0.692	0.031	0.665	0.019	0.637	0.007	0.256
27	0.952	0.487	0.000	0.487	0.000	0.487	0.000	0.239
28	1.054	0.526	0.004	0.589	0.002	0.652	0.001	0.265
29	0.952	0.487	0.000	0.487	0.000	0.487	0.000	0.239
30	1.054	0.525	0.000	0.587	0.000	0.649	0.000	0.265

TABLE 75(P).—Continued.

31	0.952	0.487	0.000	0.487	0.000	0.487	0.000	0.239
32	1.042	0.579	0.013	0.599	0.009	0.619	0.005	0.266
33	0.958	0.480	0.000	0.480	0.000	0.480	0.001	0.233
34	1.029	0.582	0.010	0.622	0.109	0.662	0.028	0.275
35	0.953	0.487	0.000	0.487	0.000	0.487	0.000	0.239
36	1.041	0.540	0.001	0.579	0.000	0.619	0.000	0.265
37	0.953	0.487	0.000	0.487	0.000	0.487	0.000	0.239
38	1.041	0.544	0.005	0.581	0.003	0.618	0.000	0.265
39	0.952	0.487	0.000	0.487	0.000	0.487	0.000	0.239
40	1.046	0.565	0.021	0.598	0.012	0.632	0.004	0.266

TABLE 75(S).—Pheophytin D.

Atom	P_{rr}	SDN_r	FOD_r	SDR_r	FRD_r	SDE_r	FED_r	π_{rr}
1	0.978	1.829	0.257	1.672	0.273	1.514	0.288	0.598
2	1.025	0.805	0.002	0.854	0.001	0.903	0.000	0.348
3	1.053	0.977	0.032	1.117	0.019	1.258	0.006	0.446
4	1.048	0.972	0.000	1.104	0.000	1.236	0.001	0.446
5	1.031	0.810	0.033	0.867	0.018	0.924	0.004	0.348
6	0.949	1.901	0.095	1.600	0.176	1.400	0.256	0.582
7	1.010	0.852	0.067	0.909	0.036	0.967	0.006	0.365
8	0.933	1.443	0.110	1.211	0.073	0.978	0.035	0.460
9	1.107	0.872	0.091	0.990	0.058	1.108	0.025	0.380
10	0.954	1.029	0.015	0.950	0.025	0.872	0.035	0.380
11	1.029	1.225	0.176	1.424	0.217	1.622	0.258	0.498
12	0.907	1.122	0.004	1.048	0.003	0.975	0.001	0.424
13	1.240	1.156	0.184	1.339	0.214	1.521	0.244	0.460
14	0.901	1.154	0.046	1.058	0.023	0.962	0.001	0.420
15	1.067	1.252	0.160	1.435	0.208	1.618	0.256	0.519
16	0.970	1.074	0.099	0.978	0.056	0.882	0.013	0.376
17	0.965	1.330	0.175	1.209	0.102	1.088	0.029	0.462
18	1.079	1.076	0.137	1.055	0.086	1.035	0.035	0.394
19	0.983	0.863	0.027	0.912	0.026	0.962	0.025	0.371
20	1.631	0.448	0.011	1.001	0.063	1.554	0.115	0.258
21	1.382	0.796	0.114	1.371	0.165	1.947	0.215	0.423
22	1.632	0.431	0.009	1.000	0.059	1.568	0.109	0.257
23	0.793	0.925	0.026	0.603	0.014	0.281	0.002	0.218
24	1.293	0.723	0.049	0.664	0.029	0.605	0.008	0.251
25	0.799	0.891	0.018	0.587	0.010	0.282	0.002	0.213
26	1.309	0.667	0.034	0.655	0.021	0.653	0.008	0.253
27	1.079	0.277	0.000	0.332	0.000	0.388	0.000	0.161
28	0.918	0.399	0.002	0.389	0.001	0.399	0.000	0.173
29	1.079	0.277	0.000	0.332	0.000	0.388	0.000	0.161
30	0.918	0.379	0.000	0.388	0.000	0.397	0.000	0.173
31	1.079	0.277	0.000	0.332	0.000	0.387	0.000	0.161
32	0.910	0.411	0.008	0.395	0.005	0.378	0.003	0.173
33	1.079	0.272	0.000	0.327	0.001	0.383	0.001	0.157
34	0.898	0.416	0.006	0.411	0.011	0.406	0.016	0.177
35	1.080	0.276	0.000	0.331	0.000	0.387	0.000	0.161
36	0.909	0.389	0.000	0.383	0.000	0.378	0.000	0.173
37	1.080	0.276	0.000	0.331	0.000	0.387	0.000	0.161
38	0.909	0.391	0.003	0.384	0.002	0.377	0.000	0.173
39	1.079	0.277	0.000	0.332	0.000	0.387	0.000	0.161
40	0.912	0.403	0.012	0.395	0.007	0.386	0.002	0.173

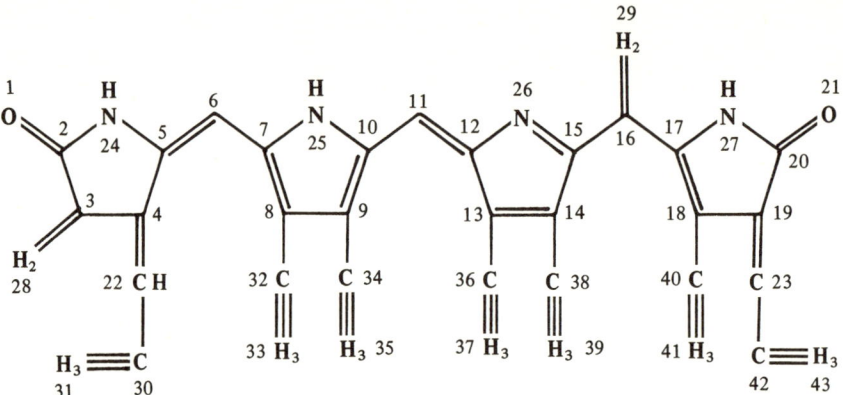

TABLE 76(P).—Phycocyanin.

Atom	P_{rr}	SDN_r	FOD_r	SDR_r	FRD_r	SDE_r	FED_r	π_{rr}
1	1.357	0.524	0.017	0.628	0.020	0.733	0.024	0.244
2	0.803	0.811	0.009	0.548	0.007	0.286	0.005	0.208
3	0.962	0.473	0.000	0.473	0.000	0.473	0.000	0.226
4	1.012	0.811	0.008	0.910	0.019	1.009	0.030	0.395
5	0.899	1.371	0.185	1.163	0.162	0.955	0.139	0.397
6	1.133	0.794	0.000	1.424	0.177	2.053	0.354	0.486
7	0.943	1.227	0.187	1.019	0.117	0.811	0.048	0.375
8	1.074	0.771	0.019	1.131	0.096	1.490	0.172	0.422
9	0.982	1.319	0.212	1.138	0.108	0.957	0.003	0.440
10	1.043	0.661	0.002	0.989	0.097	1.317	0.191	0.363
11	0.870	2.352	0.516	1.652	0.300	0.951	0.081	0.534
12	1.088	0.597	0.021	0.924	0.076	1.252	0.131	0.339
13	1.076	1.029	0.150	1.234	0.114	1.438	0.078	0.457
14	1.115	0.693	0.050	1.043	0.057	1.394	0.065	0.395
15	1.019	0.988	0.117	1.171	0.130	1.355	0.142	0.433
16	0.956	0.475	0.000	0.475	0.000	0.475	0.000	0.231
17	0.934	1.079	0.000	1.300	0.002	1.522	0.004	0.483
18	1.162	0.556	0.000	1.393	0.003	2.231	0.007	0.401
19	1.062	0.773	0.000	0.834	0.000	0.975	0.001	0.362
20	0.826	0.811	0.000	0.585	0.000	0.358	0.000	0.215
21	1.387	0.587	0.000	0.788	0.001	0.992	0.001	0.262
22	0.915	1.703	0.196	1.592	0.196	1.481	0.196	0.630
23	0.907	1.547	0.000	1.659	0.003	1.771	0.005	0.671
24	1.696	0.381	0.056	1.070	0.094	1.758	0.132	0.222
25	1.633	0.494	0.075	1.007	0.072	1.520	0.069	0.259
26	1.344	0.715	0.086	1.085	0.048	1.455	0.011	0.389
27	1.720	0.195	0.000	1.073	0.001	1.951	0.003	0.177
28	1.005	0.623	0.000	0.618	0.005	0.613	0.010	0.282
29	1.048	0.599	0.015	0.699	0.020	0.798	0.025	0.284
30	0.953	0.484	0.000	0.484	0.000	0.484	0.000	0.238
31	1.040	0.602	0.024	0.638	0.025	0.675	0.026	0.272
32	0.952	0.487	0.000	0.487	0.000	0.487	0.000	0.239
33	1.057	0.497	0.002	0.591	0.012	0.685	0.022	0.264
34	0.952	0.487	0.000	0.487	0.000	0.487	0.000	0.239
35	1.047	0.563	0.026	0.590	0.013	0.618	0.000	0.265
36	0.952	0.486	0.000	0.486	0.000	0.486	0.000	0.239

TABLE 76(P).—*Continued.*

37	1.057	0.527	0.018	0.602	0.014	0.677	0.010	0.265
38	0.951	0.487	0.000	0.487	0.000	0.487	0.000	0.239
39	1.061	0.488	0.006	0.581	0.007	0.674	0.008	0.263
40	0.952	0.486	0.000	0.486	0.000	0.486	0.000	0.239
41	1.068	0.470	0.000	0.623	0.004	0.775	0.001	0.264
42	0.954	0.484	0.000	0.484	0.000	0.484	0.000	0.238
43	1.040	0.582	0.000	0.645	0.000	0.709	0.001	0.273

TABLE 76(S).—Phycocyanin.

Atom	P_π	SDN_r	FOD_r	SDR_r	FRD_r	SDE_r	FED_r	π_π
1	1.360	0.487	0.017	0.605	0.020	0.724	0.022	0.237
2	0.794	0.800	0.009	0.543	0.007	0.285	0.005	0.208
3	1.078	0.269	0.000	0.324	0.000	0.379	0.000	0.155
4	1.018	0.771	0.006	0.906	0.019	1.040	0.032	0.394
5	0.900	1.335	0.191	1.154	0.163	0.974	0.135	0.396
6	1.139	0.774	0.000	1.449	0.173	2.124	0.345	0.484
7	0.948	1.189	0.194	1.009	0.122	0.829	0.049	0.375
8	1.077	0.746	0.017	1.145	0.094	1.544	0.170	0.422
9	0.982	1.268	0.214	1.120	0.109	0.972	0.003	0.442
10	1.049	0.650	0.003	1.006	0.096	1.363	0.189	0.362
11	0.872	2.238	0.523	1.600	0.301	0.963	0.079	0.531
12	1.096	0.586	0.025	0.943	0.079	1.299	0.133	0.339
13	1.077	1.002	0.154	1.241	0.115	1.479	0.077	0.461
14	1.122	0.667	0.051	1.052	0.057	1.437	0.064	0.494
15	1.018	0.977	0.125	1.190	0.135	1.403	0.144	0.439
16	1.077	0.271	0.000	0.328	0.001	0.385	0.001	0.157
17	0.944	1.022	0.000	1.338	0.008	1.653	0.016	0.486
18	1.169	0.540	0.000	1.469	0.002	2.397	0.028	0.402
19	1.073	0.728	0.000	0.866	0.002	1.005	0.003	0.357
20	0.827	0.801	0.000	0.585	0.001	0.369	0.001	0.215
21	1.382	0.561	0.000	0.796	0.003	1.030	0.006	0.260
22	0.915	1.651	0.205	1.599	0.201	1.546	0.196	0.638
23	0.896	1.527	0.000	1.696	0.010	1.864	0.001	0.687
24	1.697	0.363	0.057	1.067	0.091	1.771	0.125	0.219
25	1.638	0.467	0.074	1.001	0.070	1.535	0.065	0.255
26	1.355	0.665	0.082	1.073	0.045	1.480	0.009	0.380
27	1.725	0.183	0.000	1.097	0.006	2.011	0.011	0.171
28	0.889	0.423	0.000	0.402	0.003	0.382	0.005	0.179
29	0.951	0.427	0.009	0.465	0.014	0.502	0.019	0.182
30	1.079	0.276	0.000	0.331	0.000	0.387	0.001	0.160
31	0.908	0.424	0.014	0.420	0.014	0.416	0.015	0.175
32	1.079	0.276	0.000	0.333	0.000	0.389	0.001	0.161
33	0.920	0.363	0.001	0.391	0.007	0.420	0.013	0.173
34	1.079	0.277	0.000	0.332	0.000	0.387	0.000	0.161
35	0.913	0.399	0.015	0.389	0.008	0.378	0.000	0.173
36	1.079	0.277	0.000	0.333	0.000	0.389	0.000	0.161
37	0.919	0.381	0.011	0.398	0.008	0.415	0.006	0.173
38	1.079	0.277	0.000	0.333	0.000	0.389	0.000	0.161
39	0.923	0.358	0.004	0.385	0.004	0.412	0.005	0.173
40	1.079	0.276	0.000	0.334	0.000	0.391	0.000	0.161
41	0.927	0.349	0.000	0.415	0.001	0.481	0.002	0.173
42	1.079	0.275	0.000	0.331	0.000	0.387	0.000	0.161
43	0.907	0.414	0.000	0.426	0.001	0.438	0.002	0.176

TABLE 77(P).—Phycoerythrin.

Atom	P_π	SDN_r	FOD_r	SDR_r	FRD_r	SDE_r	FED_r	π_π
1	1.357	0.524	0.017	0.628	0.021	0.733	0.024	0.244
2	0.803	0.811	0.009	0.548	0.007	0.286	0.005	0.208
3	0.962	0.473	0.000	0.473	0.000	0.473	0.000	0.226
4	1.012	0.811	0.008	0.910	0.019	1.009	0.030	0.395
5	0.899	1.371	0.185	1.163	0.163	0.956	0.141	0.397
6	1.133	0.794	0.000	1.424	0.180	2.053	0.359	0.486
7	0.943	1.227	0.187	1.019	0.118	0.811	0.049	0.375
8	1.074	0.771	0.019	1.131	0.097	1.490	0.174	0.422
9	0.982	1.319	0.212	1.138	0.108	0.957	0.003	0.440
10	1.043	0.661	0.002	0.989	0.098	1.317	0.194	0.363
11	0.870	2.352	0.517	1.652	0.301	0.951	0.086	0.534
12	1.088	0.567	0.021	0.924	0.077	1.252	0.133	0.339
13	1.076	1.029	0.150	1.234	0.114	1.438	0.079	0.457
14	1.115	0.693	0.050	1.044	0.058	1.394	0.065	0.395
15	1.019	0.987	0.117	1.170	0.131	1.354	0.144	0.433
15	0.915	1.703	0.196	1.592	0.198	1.481	0.199	0.630
17	1.005	0.623	0.000	0.618	0.005	0.613	0.010	0.282
18	1.696	0.381	0.056	1.070	0.095	1.758	0.134	0.222
19	1.633	0.494	0.075	1.007	0.073	1.520	0.070	0.259
20	1.344	0.716	0.086	1.085	0.049	1.455	0.012	0.389
21	0.953	0.484	0.000	0.484	0.000	0.484	0.000	0.238
22	1.040	0.602	0.024	0.638	0.025	0.675	0.026	0.272
23	0.952	0.487	0.000	0.487	0.000	0.487	0.000	0.239
24	1.057	0.497	0.002	0.591	0.013	0.685	0.023	0.264
25	0.952	0.487	0.000	0.487	0.000	0.487	0.000	0.239
26	1.047	0.563	0.026	0.591	0.013	0.618	0.000	0.265
27	0.952	0.486	0.000	0.486	0.000	0.486	0.000	0.239
28	1.057	0.527	0.019	0.602	0.014	0.677	0.010	0.265
29	0.951	0.487	0.000	0.487	0.000	0.487	0.000	0.239
30	1.061	0.488	0.006	0.581	0.007	0.574	0.008	0.263
31	0.952	0.487	0.000	0.487	0.000	0.487	0.000	0.239
32	1.052	0.523	0.014	0.595	0.017	0.668	0.019	0.265

TABLE 77(S).—Phycoerythrin.

Atom	P_{rr}	SDN_r	FOD_r	SDR_r	FRD_r	SDE_r	FED_r	π_{rr}
1	1.360	0.487	0.017	0.605	0.020	0.724	0.023	0.237
2	0.794	0.800	0.009	0.543	0.007	0.285	0.005	0.208
3	1.078	0.269	0.000	0.324	0.000	0.379	0.000	0.155
4	1.018	0.791	0.006	0.906	0.020	1.040	0.034	0.394
5	0.900	1.335	0.191	1.154	0.166	0.974	0.141	0.196
6	1.139	0.774	0.000	1.449	0.182	2.124	0.365	0.484
7	0.948	1.189	0.194	1.009	0.122	0.828	0.051	0.385
8	1.077	0.746	0.017	1.145	0.098	1.544	0.179	0.422
9	0.892	1.268	0.214	1.120	0.109	0.972	0.003	0.442
10	1.049	0.650	0.003	1.006	0.100	1.363	0.198	0.363
11	0.872	2.238	0.523	1.600	0.304	0.962	0.086	0.530
12	1.096	0.857	0.025	0.943	0.082	1.299	0.139	0.339
13	1.077	1.002	0.154	1.241	0.117	1.480	0.081	0.461
14	1.122	0.667	0.051	1.052	0.059	1.437	0.067	0.394
15	1.018	0.977	0.125	1.190	0.138	1.403	0.150	0.438
16	0.915	1.651	0.205	1.598	0.206	1.545	0.207	0.638
17	0.889	0.423	0.000	0.402	0.003	0.382	0.006	0.179
18	1.697	0.363	0.057	1.067	0.094	1.772	0.132	0.219
19	1.638	0.467	0.074	1.001	0.072	1.535	0.069	0.255
20	1.355	0.665	0.082	1.072	0.046	1.480	0.010	0.380
21	1.079	0.276	0.000	0.331	0.001	0.387	0.001	0.160
22	0.908	0.424	0.014	0.420	0.015	0.416	0.015	0.175
23	1.079	0.276	0.000	0.333	0.000	0.389	0.000	0.161
24	0.920	0.263	0.001	0.391	0.007	0.420	0.013	0.173
25	1.079	0.277	0.000	0.332	0.000	0.387	0.000	0.161
26	0.913	0.399	0.015	0.389	0.008	0.378	0.000	0.173
27	1.079	0.277	0.000	0.333	0.000	0.389	0.000	0.161
28	0.919	0.381	0.011	0.398	0.008	0.415	0.006	0.173
29	1.079	0.277	0.000	0.333	0.000	0.389	0.000	0.161
30	0.923	0.358	0.004	0.385	0.004	0.412	0.005	0.173
31	1.080	0.277	0.000	0.332	0.000	0.388	0.001	0.161
32	0.916	0.379	0.009	0.394	0.010	0.409	0.011	0.173

TABLE 78(P).—Phytochrome (Pfr).

Atom	P_π	SDN_r	FOD_r	SDR_r	FRD_r	SDE_r	FED_r	π_{rr}
1	1.389	0.792	0.051	0.948	0.078	1.104	0.107	0.271
2	0.833	0.877	0.023	0.633	0.024	0.390	0.025	0.217
3	1.044	1.292	0.093	1.415	0.129	1.519	0.165	0.445
4	0.961	1.503	0.060	1.260	0.030	1.016	0.001	0.470
5	0.983	1.207	0.083	1.225	0.119	1.242	0.156	0.400
6	1.002	1.517	0.037	1.516	0.086	1.516	0.136	0.562
7	1.024	1.022	0.067	1.040	0.082	1.058	0.097	0.359
8	1.073	0.876	0.009	1.142	0.040	1.408	0.070	0.437
9	1.050	1.177	0.071	1.237	0.059	1.298	0.047	0.352
10	1.048	0.716	0.006	0.937	0.062	1.158	0.117	0.348
11	0.870	3.172	0.368	2.044	0.184	0.915	0.001	0.550
12	1.021	0.744	0.000	0.965	0.056	1.186	0.111	0.367
13	0.989	1.552	0.129	1.298	0.080	1.044	0.031	0.449
14	1.052	0.846	0.001	1.087	0.037	1.329	0.074	0.426
15	0.955	1.465	0.128	1.196	0.101	0.926	0.076	0.386
16	1.035	1.265	0.024	1.354	0.050	1.444	0.075	0.523
17	0.942	1.632	0.145	1.362	0.128	1.092	0.112	0.417
18	0.969	1.274	0.051	1.089	0.026	0.904	0.000	0.426
19	1.027	1.819	0.157	1.667	0.132	1.514	0.108	0.558
20	0.836	0.975	0.400	0.673	0.029	0.371	0.017	0.217
21	1.373	1.047	0.088	1.041	0.080	1.036	0.073	0.228
22	1.743	0.223	0.011	1.027	0.044	1.831	0.076	0.164
23	1.347	1.165	0.107	1.430	0.069	1.696	0.031	0.451
24	1.613	0.769	0.074	1.080	0.039	1.391	0.005	0.280
25	1.737	0.285	0.020	1.018	0.038	1.752	0.056	0.176
26	1.049	1.702	0.097	1.815	0.140	1.929	0.183	0.699
27	0.989	0.874	0.002	0.874	0.006	0.874	0.009	0.399
28	0.952	0.486	0.000	0.486	0.000	0.486	0.000	0.239
29	1.044	0.584	0.007	0.604	0.004	0.624	0.000	0.276
30	0.952	0.487	0.000	0.487	0.000	0.487	0.000	0.239
31	1.057	0.507	0.001	0.582	0.005	0.674	0.009	0.265
32	0.952	0.487	0.000	0.487	0.000	0.487	0.000	0.239
33	1.054	0.545	0.009	0.603	0.007	0.660	0.006	0.265
34	0.952	0.487	0.000	0.487	0.000	0.487	0.000	0.239
35	1.047	0.591	0.016	0.610	0.010	0.629	0.004	0.266
36	0.952	0.487	0.000	0.487	0.000	0.487	0.000	0.239
37	1.055	0.506	0.000	0.585	0.005	0.665	0.009	0.254
38	0.952	0.487	0.000	0.487	0.000	0.487	0.000	0.239
39	1.045	0.558	0.006	0.585	0.003	0.612	0.000	0.265
40	0.952	0.485	0.000	0.485	0.000	0.485	0.000	0.239
41	1.051	0.621	0.019	0.652	0.017	0.683	0.014	0.269

TABLE 78(S).—Phytochrome (Pfr).

Atom	P_π	SDN_r	FOD_r	SDR_r	FRD_r	SDE_r	FED_r	π_π
1	1.391	0.764	0.052	0.940	0.079	1.116	0.106	0.270
2	0.834	0.867	0.024	0.630	0.025	0.393	0.025	0.217
3	1.093	1.210	0.092	1.379	0.129	1.549	0.165	0.438
4	0.942	1.494	0.064	1.258	0.032	1.021	0.001	0.478
5	0.988	1.178	0.087	1.221	0.122	1.265	0.157	0.400
6	1.000	1.500	0.041	1.507	0.086	1.514	0.132	0.560
7	1.029	0.993	0.069	1.037	0.084	1.080	0.100	0.359
8	1.075	0.845	0.007	1.142	0.039	1.439	0.070	0.438
9	1.052	1.129	0.070	1.227	0.059	1.326	0.047	0.454
10	1.053	0.705	0.006	0.943	0.063	1.181	0.120	0.347
11	0.871	3.017	0.370	1.968	0.185	0.919	0.001	0.548
12	1.026	0.735	0.000	0.973	0.058	1.211	0.115	0.367
13	0.988	1.501	0.133	1.279	0.081	1.057	0.030	0.451
14	1.053	0.825	0.001	1.092	0.038	1.359	0.076	0.428
15	0.958	1.425	0.132	1.183	0.104	0.941	0.076	0.386
16	1.038	1.218	0.022	1.343	0.051	1.467	0.079	0.520
17	0.944	1.593	0.150	1.351	0.132	1.109	0.114	0.418
18	0.971	1.202	0.047	1.062	0.024	0.921	0.000	0.426
19	1.028	1.732	0.156	1.643	0.134	1.555	0.112	0.562
20	0.836	0.963	0.041	0.669	0.029	0.376	0.018	0.217
21	1.373	1.013	0.089	1.031	0.082	1.048	0.074	0.289
22	1.745	0.218	0.012	1.030	0.044	1.844	0.076	0.163
23	1.352	1.114	0.109	1.413	0.069	1.712	0.029	0.448
24	1.618	0.717	0.071	1.059	0.038	1.401	0.006	0.276
25	1.735	0.276	0.021	1.017	0.038	1.759	0.056	0.174
26	1.056	1.623	0.097	1.792	0.140	1.961	0.183	0.693
27	0.988	0.875	0.002	0.875	0.006	0.875	0.009	0.399
28	1.079	0.277	0.000	0.332	0.000	0.387	0.000	0.161
29	0.911	0.415	0.004	0.398	0.002	0.381	0.000	0.173
30	1.079	0.276	0.000	0.333	0.000	0.389	0.000	0.161
31	0.920	0.370	0.001	0.391	0.003	0.412	0.005	0.173
32	1.079	0.277	0.000	0.333	0.000	0.388	0.000	0.161
33	0.918	0.390	0.005	0.397	0.004	0.404	0.003	0.173
34	1.079	0.277	0.000	0.332	0.000	0.387	0.000	0.161
35	0.913	0.416	0.009	0.400	0.006	0.384	0.002	0.173
36	1.079	0.277	0.000	0.332	0.000	0.387	0.000	0.161
37	0.918	0.368	0.000	0.387	0.003	0.406	0.006	0.173
38	1.079	0.277	0.000	0.332	0.000	0.387	0.000	0.161
39	0.913	0.365	0.003	0.385	0.002	0.375	0.002	0.173
40	1.079	0.277	0.000	0.333	0.000	0.388	0.000	0.161
41	0.916	0.431	0.011	0.425	0.010	0.419	0.008	0.174

TABLE 79 (P).—Phytochrome (Pr).

Atom	P_{rr}	SDN_r	FOD_r	SDR_r	FRD_r	SDE_r	FED_r	π_{rr}
1	1.396	0.689	0.057	0.936	0.077	1.182	0.096	0.268
2	0.836	0.826	0.028	0.625	0.026	0.414	0.024	0.218
3	1.094	1.118	0.108	1.380	0.129	1.642	0.149	0.440
4	0.953	1.443	0.094	1.232	0.047	1.021	0.000	0.469
5	0.996	1.022	0.084	1.202	0.117	1.383	0.150	0.397
6	1.000	1.521	0.081	1.520	0.086	1.520	0.092	0.562
7	1.038	0.836	0.053	1.016	0.081	1.196	0.108	0.356
8	1.080	0.819	0.001	1.152	0.038	1.486	0.074	0.434
9	1.062	0.991	0.056	1.193	0.057	1.396	0.059	0.446
10	1.057	0.659	0.000	0.969	0.061	1.279	0.122	0.346
11	0.897	2.471	0.369	1.749	0.188	1.028	0.007	0.558
12	1.031	0.732	0.014	1.041	0.073	1.351	0.133	0.368
13	0.986	1.420	0.164	1.220	0.090	1.021	0.016	0.446
14	1.069	0.831	0.029	1.176	0.067	1.521	0.105	0.427
15	0.942	1.336	0.147	1.107	0.099	0.878	0.052	0.380
16	1.128	0.826	0.005	1.444	0.090	2.061	0.174	0.490
17	0.897	1.469	0.136	1.240	0.116	1.011	0.095	0.401
18	1.012	0.810	0.002	0.908	0.007	1.007	0.012	0.395
19	0.962	0.473	0.000	0.473	0.000	0.473	0.000	0.226
20	0.803	0.815	0.006	0.552	0.005	0.288	0.004	0.208
21	1.356	0.535	0.013	0.637	0.014	0.739	0.015	0.244
22	1.746	0.188	0.008	1.032	0.037	1.876	0.067	0.159
23	1.350	1.023	0.137	1.367	0.073	1.691	0.009	0.445
24	1.635	0.486	0.039	0.999	0.031	1.513	0.023	0.256
25	1.694	0.427	0.046	1.104	0.061	1.782	0.076	0.226
26	1.060	1.538	0.118	1.790	0.139	2.052	0.161	0.693
27	0.989	0.874	0.005	0.874	0.005	0.874	0.006	0.399
28	0.952	0.487	0.000	0.487	0.000	0.487	0.000	0.239
29	1.044	0.577	0.012	0.601	0.006	0.625	0.000	0.267
30	0.952	0.487	0.000	0.487	0.000	0.487	0.000	0.239
31	1.058	0.502	0.000	0.593	0.005	0.684	0.009	0.265
32	0.952	0.487	0.000	0.487	0.000	0.487	0.000	0.239
33	1.056	0.523	0.007	0.598	0.007	0.672	0.008	0.265
34	0.952	0.487	0.000	0.487	0.000	0.487	0.000	0.239
35	1.047	0.575	0.020	0.600	0.011	0.626	0.002	0.266
36	0.952	0.487	0.000	0.487	0.000	0.487	0.000	0.239
37	1.057	0.504	0.000	0.596	0.008	0.688	0.013	0.264
38	0.912	1.798	0.138	1.665	0.129	1.532	0.119	0.636
39	0.953	0.484	0.000	0.484	0.000	0.484	0.000	0.238
40	1.040	0.613	0.017	0.647	0.016	0.681	0.015	0.272
41	1.005	0.624	0.000	0.619	0.003	0.613	0.005	0.282

TABLE 79(S).—Phytochrome (Pr).

Atom	P_π	SDN_r	FOD_r	SDR_r	FRD_r	SDE_r	FED_r	π_π
1	1.398	0.669	0.058	0.935	0.077	1.201	0.096	0.267
2	0.836	0.828	0.029	0.624	0.027	0.419	0.025	0.218
3	1.104	1.051	0.106	1.366	0.128	1.680	0.149	0.433
4	0.944	1.436	0.099	1.232	0.050	1.027	0.000	0.474
5	1.001	1.003	0.088	1.210	0.120	1.417	0.151	0.397
6	0.998	1.503	0.087	1.510	0.087	1.517	0.087	0.560
7	1.043	0.816	0.054	1.024	0.083	1.231	0.111	0.356
8	1.081	0.797	0.001	1.158	0.037	1.520	0.073	0.435
9	1.064	0.958	0.055	1.193	0.057	1.429	0.059	0.448
10	1.061	0.652	0.000	0.982	0.062	1.313	0.124	0.345
11	0.900	2.366	0.368	1.703	0.188	1.041	0.008	0.557
12	1.036	0.727	0.016	1.058	0.078	1.389	0.136	0.368
13	0.985	1.373	0.164	1.203	0.089	1.032	0.014	0.448
14	1.070	0.080	0.028	1.188	0.068	1.567	0.108	0.428
15	0.946	1.308	0.151	1.100	0.102	0.891	0.053	0.380
16	1.132	0.810	0.006	1.467	0.093	2.124	0.181	0.488
17	0.896	1.440	0.139	1.231	0.117	1.022	0.096	0.400
18	1.018	0.770	0.001	0.904	0.008	1.038	0.014	0.393
19	1.078	0.269	0.000	0.325	0.000	0.479	0.000	0.155
20	0.794	0.804	0.006	0.546	0.005	0.287	0.004	0.208
21	1.359	0.498	0.013	0.613	0.013	0.728	0.014	0.237
22	1.748	0.184	0.009	1.037	0.037	1.891	0.066	0.158
23	1.355	0.983	0.138	1.345	0.073	1.708	0.008	0.442
24	1.640	0.459	0.037	0.993	0.031	1.527	0.024	0.252
25	1.693	0.411	0.047	1.100	0.061	1.790	0.076	0.224
26	1.067	1.464	0.117	1.778	0.139	2.093	0.160	0.687
27	0.988	0.875	0.005	0.875	0.006	0.875	0.006	0.339
28	1.079	0.277	0.000	0.332	0.000	0.387	0.000	0.161
29	0.911	0.411	0.007	0.396	0.003	0.382	0.000	0.173
30	1.079	0.276	0.000	0.333	0.000	0.389	0.000	0.161
31	0.920	0.366	0.000	0.392	0.003	0.418	0.005	0.173
32	1.079	0.277	0.000	0.332	0.000	0.387	0.000	0.161
33	0.919	0.378	0.004	0.394	0.004	0.411	0.004	0.173
34	1.079	0.277	0.000	0.333	0.000	0.389	0.000	0.161
35	0.913	0.407	0.012	0.395	0.006	0.383	0.001	0.173
36	1.079	0.276	0.000	0.333	0.000	0.389	0.000	0.161
37	0.919	0.367	0.002	0.394	0.005	0.421	0.008	0.173
38	0.910	1.757	0.143	1.674	0.134	1.590	0.125	0.644
39	1.079	0.276	0.000	0.331	0.000	0.387	0.000	0.161
40	0.908	0.431	0.010	0.425	0.010	0.419	0.009	0.175
41	0.889	0.423	0.000	0.403	0.002	0.382	0.003	0.179

TABLE 80.—Pimpinellin.

Atom	P_{rr}	SDN_r	FOD_r	SDR_r	FRD_r	SDE_r	FED_r	π_{rr}
1	1.114	0.653	0.014	0.967	0.110	1.280	0.205	0.402
2	0.994	0.915	0.001	1.148	0.140	1.380	0.279	0.491
3	1.791	0.199	0.023	0.586	0.051	0.973	0.078	0.115
4	1.005	0.796	0.176	0.897	0.088	0.998	0.001	0.391
5	1.055	0.661	0.001	1.034	0.178	1.408	0.355	0.392
6	0.996	0.842	0.168	1.074	0.237	1.305	0.305	0.422
7	1.098	0.566	0.029	0.785	0.020	1.005	0.012	0.332
8	0.909	1.193	0.585	0.964	0.283	0.734	0.001	0.420
9	1.097	0.917	0.449	1.137	0.229	1.357	0.009	0.487
10	0.814	0.758	0.107	0.529	0.056	0.299	0.005	0.208
11	1.829	0.174	0.073	0.589	0.074	1.004	0.076	0.097
12	0.949	0.958	0.217	1.094	0.267	1.229	0.317	0.433
13	1.093	0.559	0.012	0.792	0.069	1.025	0.126	0.329
14	1.937	0.047	0.000	0.623	0.055	1.199	0.110	0.037
15	1.925	0.072	0.019	0.619	0.057	1.166	0.094	0.047
16	1.403	0.461	0.145	0.657	0.076	0.852	0.028	0.239

TABLE 81(P).—Plastoquinone.

Atom	$P_{\pi\pi}$	SDN_r	FOD_r	SDR_r	FRD_r	SDE_r	FED_r	$\pi_{\pi\pi}$
1	0.952	1.614	0.221	1.282	0.119	0.949	0.017	0.474
2	0.949	1.641	0.227	1.293	0.122	0.946	0.016	0.476
3	0.826	1.234	0.160	0.754	0.080	0.276	0.000	0.206
4	0.999	1.663	0.233	1.323	0.145	0.984	0.058	0.502
5	0.907	1.783	0.227	1.303	0.142	0.823	0.027	0.458
6	0.823	1.263	0.166	0.768	0.083	0.274	0.000	0.207
7	1.274	1.703	0.330	1.163	0.165	0.623	0.000	0.298
8	0.955	0.474	0.000	0.474	0.012	0.474	0.024	0.229
9	0.993	0.984	0.000	1.085	0.364	1.186	0.728	0.511
10	1.000	0.907	0.000	1.008	0.334	1.109	0.668	0.461
11	0.952	0.486	0.000	0.486	0.000	0.486	0.000	0.239
12	1.043	0.597	0.027	0.606	0.015	0.615	0.003	0.267
13	0.952	0.486	0.000	0.486	0.000	0.486	0.000	0.239
14	1.043	0.597	0.027	0.606	0.015	0.615	0.003	0.267
15	0.953	0.486	0.000	0.486	0.008	0.486	0.016	0.239
16	1.050	0.511	0.000	0.573	0.059	0.636	0.119	0.265
17	0.953	0.486	0.000	0.486	0.008	0.486	0.016	0.239
18	1.050	0.511	0.000	0.573	0.059	0.636	0.119	0.265
19	1.288	1.645	0.318	1.146	0.160	0.647	0.003	0.298
20	1.038	0.679	0.032	0.863	0.108	0.687	0.183	0.287

TABLE 81(S).—Plastoquinone.

Atom	$P_{\pi\pi}$	SDN_r	FOD_r	SDR_r	FRD_r	SDE_r	FED_r	$\pi_{\pi\pi}$
1	0.956	1.511	0.215	1.240	0.117	0.990	0.019	0.475
2	0.953	1.547	0.225	1.256	0.122	0.966	0.018	0.477
3	0.825	1.200	0.166	0.737	0.083	0.775	0.000	0.206
4	1.011	1.552	0.232	1.275	0.247	0.998	0.062	0.496
5	0.899	1.753	0.272	1.289	0.151	0.827	0.030	0.462
6	0.824	1.238	0.166	0.756	0.088	0.275	0.000	0.207
7	1.274	1.636	0.341	1.131	0.171	0.626	0.000	0.299
8	1.076	0.271	0.000	0.326	0.009	0.381	0.018	0.156
9	1.017	0.896	0.000	1.071	0.412	1.247	0.824	0.511
10	0.999	0.876	0.000	1.014	0.381	1.152	0.761	0.467
11	1.079	0.277	0.000	0.332	0.000	0.387	0.000	0.161
12	0.911	0.416	0.015	0.397	0.008	0.377	0.002	0.173
13	1.079	0.277	0.000	0.332	0.000	0.387	0.000	0.161
14	0.911	0.416	0.015	0.398	0.008	0.377	0.002	0.173
15	1.079	0.276	0.000	0.332	0.006	0.387	0.012	0.160
16	0.915	0.371	0.000	0.381	0.034	0.391	0.067	0.173
17	1.079	0.276	0.000	0.332	0.006	0.387	0.012	0.160
18	0.915	0.371	0.000	0.381	0.034	0.391	0.067	0.173
19	1.290	1.554	0.322	1.102	0.162	0.651	0.003	0.298
20	0.906	0.470	0.019	0.450	0.061	0.429	0.103	0.183

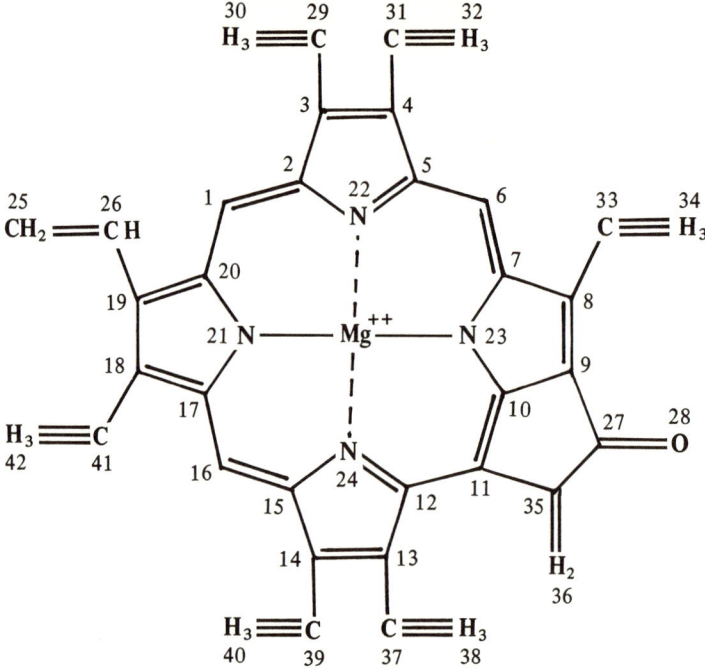

TABLE 82(P).—Protochlorophyll A.

Atom	P_{π}	SDN_r	FOD_r	SDR_r	FRD_r	SDE_r	FED_r	π_π
1	0.963	2.561	0.247	1.795	0.218	1.029	0.188	0.528
2	0.914	1.486	0.003	1.133	0.011	0.781	0.019	0.385
3	0.981	1.628	0.106	1.298	0.082	0.967	0.057	0.439
4	0.972	1.553	0.001	1.248	0.039	0.943	0.078	0.430
5	0.925	1.558	0.108	1.180	0.057	0.803	0.006	0.386
6	0.941	2.439	0.027	1.721	0.113	1.003	0.199	0.533
7	0.929	1.554	0.123	1.177	0.069	0.800	0.014	0.384
8	0.920	1.887	0.063	1.374	0.055	0.861	0.047	0.447
9	1.048	1.542	0.132	1.232	0.097	0.922	0.061	0.398
10	0.904	1.472	0.004	1.125	0.015	0.779	0.026	0.391
11	0.917	2.627	0.256	1.825	0.250	1.024	0.244	0.515
12	0.922	1.459	0.002	1.123	0.009	0.788	0.017	0.388
13	0.974	1.670	0.112	1.310	0.097	0.947	0.082	0.437
14	0.976	1.528	0.001	1.243	0.044	0.959	0.088	0.439
15	0.919	1.598	0.113	1.187	0.063	0.775	0.012	0.383
16	0.952	2.370	0.024	1.713	0.136	1.056	0.248	0.540
17	0.920	1.592	0.128	1.181	0.069	0.770	0.011	0.381
18	0.962	1.736	0.048	1.384	0.086	1.032	0.125	0.472
19	1.008	1.565	0.136	1.215	0.117	0.866	0.098	0.379
20	0.902	1.508	0.014	1.156	0.008	0.804	0.003	0.396
21	1.758	0.440	0.012	0.662	0.011	0.884	0.010	0.133
22	1.759	0.443	0.031	0.656	0.025	0.869	0.018	0.130
23	1.749	0.489	0.018	0.677	0.025	0.864	0.032	0.137
24	1.760	0.439	0.031	0.658	0.027	0.876	0.023	0.130
25	0.987	2.025	0.141	1.676	0.145	1.326	0.149	0.647
26	0.994	0.893	0.003	0.893	0.015	0.893	0.028	0.408
27	0.798	0.976	0.018	0.625	0.010	0.274	0.002	0.212
28	1.289	0.876	0.041	0.735	0.029	0.594	0.017	0.257
29	0.952	0.487	0.000	0.487	0.000	0.487	0.000	0.239

TABLE 82(P).—*Continued.*

30	1.046	0.601	0.013	0.610	0.011	0.620	0.008	0.265
31	0.952	0.487	0.000	0.487	0.000	0.487	0.000	0.239
32	1.045	0.591	0.000	0.604	0.006	0.617	0.011	0.266
33	0.952	0.487	0.000	0.487	0.000	0.487	0.000	0.239
34	1.040	0.631	0.008	0.618	0.007	0.605	0.007	0.266
35	0.958	0.480	0.000	0.480	0.001	0.480	0.001	0.233
36	1.018	0.684	0.017	0.643	0.022	0.603	0.028	0.275
37	0.952	0.487	0.000	0.487	0.000	0.487	0.001	0.239
38	1.046	0.606	0.014	0.612	0.013	0.617	0.012	0.265
39	0.952	0.487	0.000	0.487	0.000	0.487	0.001	0.239
40	1.046	0.588	0.000	0.603	0.006	0.619	0.012	0.265
41	0.952	0.486	0.000	0.486	0.000	0.486	0.001	0.239
42	1.044	0.613	0.006	0.619	0.012	0.626	0.018	0.267

TABLE 82(S).—Protochlorophyll *A*.

Atom	P_π	SDN_r	FOD_r	SDR_r	FRD_r	SDE_r	FED_r	π_π
1	0.966	2.426	0.249	1.734	0.222	1.042	0.194	0.527
2	0.918	1.435	0.002	1.110	0.010	0.786	0.018	0.385
3	0.981	1.555	0.110	1.270	0.086	0.895	0.061	0.441
4	0.971	1.483	0.001	1.220	0.040	0.957	0.079	0.440
5	0.929	1.562	0.110	1.156	0.058	0.811	0.006	0.386
6	0.941	2.314	0.025	1.663	0.114	1.011	0.203	0.532
7	0.933	1.497	0.125	1.151	0.070	0.806	0.013	0.383
8	0.913	1.817	0.061	1.343	0.056	0.869	0.050	0.451
9	1.053	1.446	0.130	1.185	0.096	0.925	0.061	0.384
10	0.910	1.414	0.003	1.101	0.016	0.789	0.028	0.390
11	0.909	2.508	0.259	1.771	0.254	1.034	0.249	0.518
12	0.929	1.402	0.002	1.098	0.009	0.794	0.016	0.387
13	0.973	1.593	0.114	1.278	0.100	0.963	0.085	0.440
14	0.976	1.454	0.001	1.215	0.046	0.957	0.091	0.441
15	0.923	1.537	0.115	1.160	0.064	0.782	0.013	0.383
16	0.952	2.257	0.026	1.661	0.138	1.065	0.251	0.540
17	0.926	1.533	0.131	1.155	0.071	0.777	0.011	0.380
18	0.954	1.690	0.050	1.366	0.088	1.042	0.127	0.477
19	1.014	1.472	0.135	1.174	0.117	0.877	0.100	0.377
20	0.904	1.457	0.013	1.133	0.008	0.809	0.003	0.397
21	1.759	0.419	0.013	0.653	0.012	0.887	0.010	0.132
22	1.761	0.414	0.030	0.643	0.024	0.872	0.018	0.129
23	1.751	0.473	0.020	0.670	0.026	0.867	0.032	0.137
24	1.762	0.416	0.031	0.647	0.027	0.879	0.023	0.129
25	0.992	1.935	0.141	1.637	0.145	1.340	0.150	0.645
26	0.994	0.894	0.003	0.894	0.015	0.894	0.028	0.408
27	0.795	0.965	0.020	0.619	0.011	0.274	0.002	0.212
28	1.294	0.851	0.044	0.723	0.031	0.596	0.018	0.255
29	1.079	0.277	0.000	0.332	0.000	0.387	0.000	0.161
30	0.913	0.420	0.008	0.399	0.006	0.379	0.005	0.173
31	1.079	0.277	0.000	0.332	0.000	0.387	0.000	0.161
32	0.912	0.415	0.000	0.396	0.003	0.377	0.006	0.173
33	1.079	0.277	0.000	0.332	0.000	0.386	0.000	0.161
34	0.908	0.438	0.004	0.404	0.004	0.370	0.004	0.173
35	1.079	0.272	0.000	0.327	0.001	0.381	0.001	0.157
36	0.892	0.469	0.010	0.420	0.013	0.371	0.016	0.177
37	1.079	0.277	0.000	0.332	0.000	0.387	0.001	0.161
38	0.912	0.422	0.008	0.400	0.007	0.378	0.007	0.173
39	1.079	0.277	0.000	0.332	0.000	0.387	0.001	0.161
40	0.913	0.413	0.000	0.396	0.004	0.378	0.007	0.173
41	1.079	0.277	0.000	0.332	0.001	0.387	0.001	0.161
42	0.911	0.429	0.003	0.406	0.007	0.383	0.010	0.173

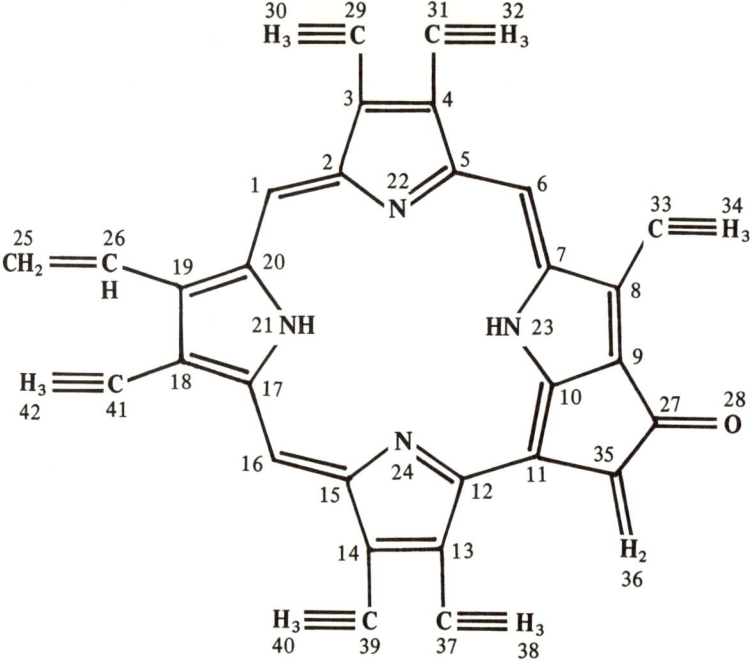

TABLE 83(P).—Protopheophytin A.

Atom	P_{rr}	SDN_r	FOD_r	SDR_r	FRD_r	SDE_r	FED_r	π_{rr}
1	0.970	1.970	0.229	1.713	0.241	1.457	0.254	0.589
2	1.007	0.989	0.064	0.907	0.032	0.825	0.001	0.352
3	1.040	1.165	0.118	1.174	0.060	1.183	0.003	0.449
4	1.036	1.170	0.089	1.171	0.046	1.172	0.002	0.451
5	1.012	0.982	0.092	0.908	0.046	0.834	0.001	0.350
6	0.944	1.003	0.078	1.703	0.164	1.403	0.243	0.594
7	0.975	0.998	0.060	0.923	0.035	0.849	0.010	0.366
8	0.943	1.413	0.002	1.183	0.016	0.954	0.029	0.451
9	1.076	1.073	0.067	1.047	0.043	1.020	0.029	0.396
10	0.969	1.046	0.034	0.958	0.032	0.870	0.029	0.379
11	0.916	2.038	0.238	1.726	0.253	1.413	0.268	0.570
12	1.013	0.976	0.061	0.901	0.031	0.827	0.001	0.353
13	1.037	1.176	0.118	1.172	0.060	1.167	0.002	0.449
14	1.038	1.158	0.086	1.167	0.045	1.176	0.003	0.450
15	1.009	0.992	0.092	0.904	0.046	0.816	0.000	0.350
16	0.956	1.952	0.070	1.715	0.173	1.478	0.275	0.601
17	0.978	1.015	0.062	0.927	0.038	0.839	0.014	0.365
18	0.986	1.328	0.002	1.246	0.028	1.164	0.054	0.477
19	1.037	1.073	0.058	1.035	0.052	0.997	0.046	0.379
20	0.965	1.036	0.023	0.956	0.018	0.872	0.013	0.379
21	1.625	0.637	0.089	1.101	0.086	1.565	0.083	0.278
22	1.378	0.781	0.007	1.354	0.103	1.928	0.198	0.422
23	1.610	0.730	0.102	1.114	0.107	1.525	0.112	0.289
24	1.376	0.790	0.009	1.368	0.111	1.945	0.214	0.424
25	1.015	1.535	0.064	1.497	0.059	1.459	0.054	0.641
26	0.993	0.894	0.003	0.894	0.003	0.894	0.004	0.407
27	0.807	0.919	0.009	0.597	0.005	0.276	0.002	0.208
28	1.304	0.754	0.017	0.693	0.013	0.632	0.009	0.262
29	0.952	0.487	0.000	0.487	0.000	0.487	0.000	0.239

TABLE 83(P).—*Continued.*

30	1.053	0.544	0.014	0.595	0.007	0.646	0.000	0.265
31	0.952	0.487	0.000	0.487	0.000	0.487	0.000	0.239
32	1.053	0.544	0.014	0.595	0.007	0.646	0.000	0.265
33	0.952	0.487	0.000	0.487	0.000	0.487	0.000	0.239
34	1.043	0.573	0.000	0.595	0.002	0.617	0.004	0.266
35	0.962	0.473	0.000	0.473	0.000	0.473	0.000	0.226
36	1.000	0.691	0.015	0.660	0.021	0.629	0.027	0.284
37	0.952	0.487	0.000	0.487	0.000	0.487	0.000	0.239
38	1.053	0.544	0.014	0.595	0.007	0.644	0.000	0.265
39	0.952	0.487	0.000	0.487	0.000	0.487	0.000	0.239
40	1.053	0.544	0.014	0.595	0.007	0.644	0.000	0.265
41	0.952	0.487	0.000	0.487	0.000	0.487	0.000	0.239
42	1.047	0.563	0.000	0.602	0.004	0.642	0.007	0.267

TABLE 83(S).—Protopheophytin *A*.

Atom	P_{rr}	SDN_r	FOD_r	SDR_r	FRD_r	SDE_r	FED_r	π_{rr}
1	0.975	1.895	0.240	1.689	0.249	1.484	0.258	0.587
2	1.009	0.975	0.064	0.903	0.032	0.829	0.000	0.352
3	1.043	1.116	0.116	1.164	0.059	1.212	0.003	0.450
4	1.038	1.123	0.083	1.161	0.042	1.198	0.002	0.453
5	1.015	0.966	0.097	0.903	0.049	0.840	0.001	0.350
6	0.945	1.941	0.069	1.680	0.160	1.419	0.251	0.592
7	0.989	0.978	0.064	0.915	0.037	0.853	0.009	0.365
8	0.935	1.398	0.001	1.181	0.016	0.964	0.031	0.456
9	1.082	1.023	0.062	1.003	0.046	1.023	0.029	0.390
10	0.973	1.021	0.036	0.952	0.033	0.883	0.031	0.377
11	0.907	1.895	0.251	1.704	0.260	1.424	0.268	0.570
12	1.018	0.953	0.059	0.893	0.030	0.833	0.002	0.352
13	1.039	1.124	0.115	1.159	0.058	1.194	0.002	0.050
14	1.041	1.103	0.077	1.159	0.040	1.244	0.004	0.451
15	1.012	0.972	0.092	0.897	0.048	0.821	0.000	0.349
16	0.957	1.888	0.057	1.691	0.167	1.494	0.277	0.599
17	0.982	0.996	0.066	0.921	0.040	0.845	0.014	0.365
18	0.978	1.322	0.001	1.249	0.028	1.175	0.056	0.483
19	1.044	1.028	0.065	1.020	0.057	1.013	0.048	0.376
20	0.966	1.023	0.021	0.950	0.017	0.876	0.013	0.379
21	1.628	0.609	0.088	1.092	0.085	1.575	0.082	0.275
22	1.384	0.752	0.009	1.354	0.104	1.957	0.198	0.418
23	1.611	0.686	0.106	1.110	0.109	1.534	0.112	0.288
24	1.380	0.771	0.012	1.369	0.112	1.968	0.212	0.422
25	1.020	1.492	0.072	1.485	0.064	1.477	0.056	0.639
26	0.993	0.895	0.004	0.895	0.004	0.895	0.004	0.408
27	0.800	0.908	0.012	0.592	0.007	0.276	0.002	0.209
28	1.307	0.720	0.023	0.674	0.017	0.629	0.010	0.255
29	1.079	0.277	0.000	0.332	0.000	0.388	0.000	0.161
30	0.917	0.389	0.008	0.392	0.004	0.396	0.000	0.173
31	1.079	0.277	0.000	0.332	0.000	0.388	0.000	0.161
32	0.917	0.389	0.008	0.392	0.004	0.396	0.000	0.173
33	1.079	0.277	0.000	0.322	0.000	0.388	0.000	0.161
34	0.910	0.408	0.000	0.393	0.001	0.377	0.002	0.173
35	1.078	0.269	0.000	0.324	0.000	0.378	0.000	0.155
36	0.884	0.465	0.009	0.428	0.013	0.391	0.016	0.180
37	1.079	0.277	0.000	0.332	0.000	0.388	0.000	0.161
38	0.917	0.389	0.008	0.392	0.004	0.396	0.000	0.173
39	1.079	0.277	0.000	0.332	0.000	0.388	0.001	0.161
40	0.917	0.389	0.008	0.392	0.004	0.396	0.000	0.173
41	1.079	0.277	0.000	0.332	0.000	0.388	0.000	0.161
42	0.913	0.403	0.000	0.397	0.003	0.392	0.004	0.174

TABLE 84.—Psoralen.

Atom	P_π	SDN_r	FOD_r	SDR_r	FRD_r	SDE_r	FED_r	π_{rr}
1	1.073	1.011	0.418	1.133	0.220	1.155	0.023	0.493
2	0.910	1.237	0.487	0.991	0.244	0.744	0.000	0.425
3	1.051	0.631	0.069	0.753	0.040	0.875	0.020	0.332
4	0.996	1.113	0.392	1.118	0.302	1.123	0.212	0.487
5	1.055	0.614	0.014	0.736	0.083	0.858	0.151	0.327
6	1.122	0.651	0.032	0.947	0.183	1.243	0.335	0.402
7	0.972	0.978	0.049	1.100	0.268	1.222	0.488	0.490
8	1.792	0.192	0.003	0.558	0.028	0.923	0.052	0.114
9	0.967	0.854	0.114	0.874	0.125	0.893	0.107	0.382
10	1.087	0.753	0.083	0.944	0.130	0.135	0.176	0.420
11	0.940	0.893	0.045	0.913	0.174	0.932	0.304	0.390
12	1.835	0.159	0.035	0.564	0.062	0.971	0.089	0.091
13	0.810	0.779	0.105	0.533	0.255	0.286	0.004	0.206
14	1.392	0.501	0.133	0.650	0.086	0.798	0.039	0.242

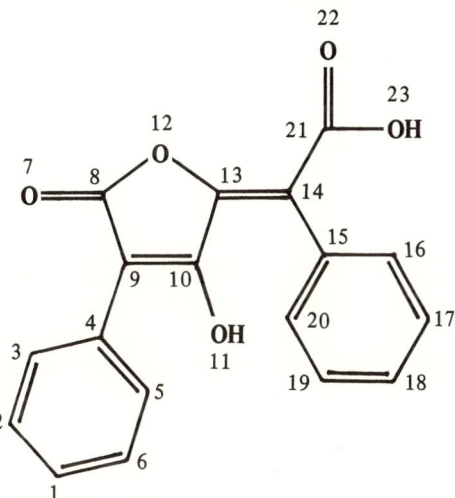

TABLE 85.—Pulvinic Acid.

Atom	P_{rr}	SDN_r	FOD_r	SDR_r	FRD_r	SDE_r	FED_r	π_{rr}
1	1.014	0.955	0.072	0.992	0.111	1.030	0.750	0.422
2	0.999	0.828	0.001	0.828	0.004	0.828	0.006	0.395
3	1.017	0.980	0.066	1.018	0.097	1.055	0.127	0.439
4	0.991	0.749	0.013	0.749	0.030	0.749	0.049	0.335
5	1.017	0.980	0.066	1.018	0.097	1.055	0.127	0.439
6	0.999	0.828	0.001	0.828	0.004	0.828	0.006	0.395
7	1.357	0.750	0.123	0.777	0.121	0.803	0.119	0.260
8	0.807	0.888	0.068	0.597	0.043	0.307	0.019	0.210
9	1.108	1.133	0.233	1.284	0.311	1.435	0.389	0.450
10	0.902	1.455	0.261	1.165	0.186	0.874	0.111	0.423
11	1.907	0.178	0.041	0.619	0.038	1.060	0.035	0.064
12	1.850	0.123	0.010	0.547	0.010	0.971	0.011	0.077
13	0.910	1.460	0.068	1.170	0.186	0.879	0.104	0.426
14	1.039	1.439	0.323	1.341	0.316	1.243	0.308	0.474
15	0.994	0.747	0.018	0.747	0.028	0.747	0.039	0.334
16	1.004	1.055	0.092	1.030	0.096	1.005	0.100	0.441
17	1.000	0.828	0.002	0.828	0.003	0.828	0.005	0.395
18	1.002	1.029	1.000	1.004	0.109	0.980	0.119	0.424
19	1.000	0.828	0.002	0.828	0.003	0.828	0.005	0.395
20	1.004	1.055	0.092	1.030	0.096	1.005	0.100	0.441
21	0.795	0.860	0.049	0.570	0.029	0.279	0.009	0.205
22	1.358	0.663	0.089	0.690	0.074	0.716	0.059	0.247
23	1.924	0.069	0.008	0.511	0.005	0.952	0.003	0.042

TABLE 86.—Purine.

Atom	P_π	SDN_r	FOD_r	SDR_r	FRD_r	SDE_r	FED_r	π_π
1	1.204	0.795	0.409	0.883	0.284	0.971	0.159	0.395
2	0.905	0.997	0.190	0.886	0.267	0.774	0.283	0.404
3	1.193	0.727	0.101	0.815	0.051	0.903	0.001	0.376
4	0.910	1.138	0.636	0.970	0.465	0.801	0.293	0.432
5	1.038	0.639	0.074	0.727	0.095	0.816	0.116	0.331
6	1.313	0.537	0.093	0.889	0.308	1.242	0.524	0.358
7	0.897	1.067	0.383	0.952	0.343	0.838	0.302	0.435
8	1.592	0.347	0.029	0.783	0.088	1.218	0.147	0.247
9	0.948	0.810	0.085	0.783	0.130	0.756	0.174	0.356

TABLE 87(P).—Pyridoxal.

Atom	P_{π}	SDN_r	FOD_r	SDR_r	FRD_r	SDE_r	FED_r	π_{π}
1	1.149	0.924	0.448	0.873	0.224	0.822	0.000	0.395
2	0.959	0.874	0.205	0.890	0.337	0.907	0.468	0.399
3	0.924	0.917	0.137	0.860	0.270	0.803	0.409	0.381
4	1.047	0.799	0.347	0.828	0.192	0.857	0.367	0.367
5	0.944	0.929	0.242	0.878	0.253	0.826	0.265	0.396
6	0.990	0.860	0.070	0.892	0.266	0.923	0.462	0.414
7	0.789	0.904	0.219	0.588	0.109	0.272	0.000	0.216
8	1.285	0.653	0.248	0.608	0.128	0.563	0.008	0.249
9	1.918	0.078	0.016	0.567	0.111	1.055	0.206	0.051
10	0.953	0.488	0.003	0.488	0.007	0.488	0.011	0.239
11	1.046	0.510	0.030	0.562	0.057	0.614	0.084	0.264
12	0.953	0.488	0.004	0.488	0.005	0.488	0.006	0.239
13	1.044	0.516	0.035	0.560	0.041	0.604	0.047	0.264

TABLE 87(S).—Pyridoxal.

Atom	P_{π}	SDN_r	FOD_r	SDR_r	FRD_r	SDE_r	FED_r	π_{π}
1	1.153	0.906	0.456	0.865	0.228	0.824	0.000	0.393
2	0.955	0.864	0.181	0.891	0.332	0.917	0.483	0.401
3	0.929	0.508	0.176	0.859	0.299	0.811	0.421	0.382
4	1.051	0.782	0.345	0.820	0.194	0.859	0.042	0.365
5	0.940	0.918	0.214	0.877	0.242	0.836	0.269	0.398
6	0.996	0.848	0.090	0.890	0.288	0.933	0.481	0.413
7	0.789	0.902	0.228	0.587	0.114	0.272	0.000	0.216
8	1.285	0.648	0.256	0.606	0.133	0.563	0.009	0.249
9	1.918	0.077	0.020	0.508	0.118	1.058	0.216	0.050
10	1.080	0.267	0.000	0.332	0.004	0.387	0.007	0.160
11	0.912	0.371	0.031	0.373	0.028	0.375	0.043	0.172
12	1.080	0.276	0.000	0.332	0.002	0.387	0.004	0.161
13	0.911	0.375	0.016	0.372	0.020	0.369	0.024	0.172

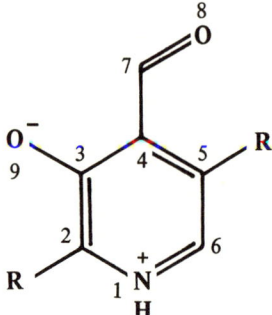

TABLE 88(I).—Pyridoxal-1.

Atom	$P_{\pi\pi}$	SDN_r	FOD_r	SDR_r	FRD_r	SDE_r	FED_r	π_{rr}
1	1.207	0.785	0.432	0.848	0.235	0.911	0.038	0.382
2	0.831	0.788	0.079	0.920	0.311	1.053	0.543	0.397
3	0.978	0.866	0.346	0.916	0.367	0.967	0.388	0.397
4	1.099	0.677	0.297	0.801	0.149	0.916	0.000	0.354
5	0.817	0.832	0.094	0.895	0.239	0.957	0.384	0.396
6	1.061	0.766	0.221	0.955	0.353	1.135	0.484	0.422
7	0.790	0.885	0.240	0.579	0.120	0.272	0.000	0.216
8	1.294	0.613	0.254	0.595	0.127	0.576	0.000	0.247
9	1.923	0.075	0.037	0.585	0.100	2.095	0.163	0.048

TABLE 89(I).—Pyridoxal-2 (zwitterion).

Atom	$P_{\pi\pi}$	SDN_r	FOD_r	SDR_r	FRD_r	SDE_r	FED_r	π_{rr}
1	1.650	0.598	0.248	0.651	0.133	0.704	0.015	0.182
2	0.578	1.582	0.242	1.156	0.341	0.729	0.247	0.391
3	0.536	1.057	0.273	0.894	0.198	0.732	0.123	0.362
4	1.069	0.967	0.320	1.027	0.272	1.086	0.223	0.405
5	0.769	0.856	0.004	0.800	0.027	0.744	0.050	0.366
6	1.163	1.072	0.323	1.294	0.396	1.515	0.469	0.527
7	0.794	0.912	0.106	0.319	0.056	0.286	0.007	0.219
8	1.292	0.692	0.153	0.655	0.102	0.619	0.051	0.253
9	1.758	0.365	0.139	0.409	0.476	2.453	0.814	0.264

TABLE 90(I).—Pyridoxal, anionic.

Atom	P_π	SDN_r	FOD_r	SDR_r	FRD_r	SDE_r	FED_r	π_π
1	1.213	0.754	0.432	0.844	0.224	0.935	0.017	0.379
2	0.844	0.793	0.130	1.096	0.261	1.398	0.392	0.419
3	0.967	0.814	0.309	0.866	0.241	0.918	0.173	0.362
4	1.139	0.609	0.259	0.889	0.179	1.170	0.099	0.358
5	0.819	0.828	0.110	0.918	0.121	1.009	0.132	0.396
6	1.100	0.706	0.173	1.114	0.285	1.523	0.397	0.428
7	0.795	0.866	0.241	0.576	0.122	0.285	0.004	0.218
8	1.306	0.582	0.245	0.609	0.134	0.635	0.023	0.248
9	1.808	0.187	0.100	1.503	0.342	2.819	0.763	0.182

TABLE 91(P).—Pyridoxal, Schiff-Base.

Atom	P_{rr}	SDN_r	FOD_r	SDR_r	FRD_r	SDE_r	FED_r	π_{rr}
1	1.152	1.006	0.276	0.936	0.186	0.865	0.097	0.404
2	0.961	0.881	0.088	0.895	0.112	0.909	0.135	0.399
3	0.924	1.015	0.157	0.938	0.280	0.860	0.403	0.399
4	1.032	0.760	0.126	0.789	0.167	0.818	0.208	0.344
5	0.950	1.023	0.202	0.953	0.101	0.882	0.000	0.413
6	0.990	0.858	0.026	0.889	0.117	0.920	0.208	0.413
7	0.851	1.454	0.459	1.059	0.282	0.664	0.105	0.420
8	1.219	1.156	0.456	1.185	0.457	1.214	0.458	0.492
9	0.956	0.459	0.008	0.459	0.013	0.459	0.018	0.216
10	0.775	0.818	0.002	0.535	0.001	0.252	0.000	0.202
11	1.335	0.484	0.003	0.532	0.003	0.580	0.003	0.233
12	1.921	0.057	0.000	0.499	0.000	0.942	0.000	0.043
13	1.024	0.750	0.135	0.735	0.145	0.721	0.152	0.309
14	1.915	0.096	0.021	0.580	0.104	1.065	0.188	0.054
15	0.953	0.487	0.001	0.487	0.001	0.487	0.001	0.239
16	1.011	0.527	0.026	0.569	0.013	0.610	0.000	0.264
17	0.953	0.488	0.001	0.488	0.002	0.488	0.003	0.239
18	1.043	0.511	0.012	0.563	0.017	0.614	0.023	0.264

TABLE 91(S).—Pyridoxal, Schiff-Base.

Atom	P_{rr}	SDN_r	FOD_r	SDR_r	FRD_r	SDE_r	FED_r	π_{rr}
1	1.158	0.963	0.295	0.916	0.189	0.869	0.084	0.401
2	0.957	0.868	0.090	0.894	0.137	0.920	0.184	0.401
3	0.933	0.981	0.175	0.926	0.312	0.871	0.450	0.398
4	1.035	0.748	0.145	0.786	0.175	0.825	0.206	0.344
5	0.948	0.987	0.195	0.940	0.100	0.893	0.005	0.414
6	0.996	0.845	0.039	0.887	0.152	0.930	0.265	0.412
7	0.862	1.340	0.437	0.999	0.261	0.658	0.082	0.411
8	1.217	1.175	0.522	1.213	0.474	1.251	0.426	0.521
9	1.072	0.267	0.000	0.322	0.004	0.377	0.009	0.153
10	0.767	0.808	0.000	0.530	0.000	0.252	0.000	0.202
11	1.339	0.451	0.000	0.512	0.001	0.571	0.002	0.227
12	1.921	0.054	0.000	0.498	0.000	0.941	0.000	0.043
13	0.897	0.480	0.058	0.453	0.057	0.426	0.056	0.188
14	1.917	0.091	0.022	0.580	0.118	1.069	0.214	0.053
15	1.080	0.276	0.000	0.332	0.000	0.387	0.000	0.161
16	0.911	0.379	0.014	0.376	0.007	0.373	0.000	0.173
17	1.080	0.276	0.000	0.332	0.001	0.387	0.003	0.161
18	0.912	0.371	0.006	0.371	0.011	0.375	0.016	0.172

TABLE 92(P).—Pyridoxine.

Atom	P_π	SDN_r	FOD_r	SDR_r	FRD_r	SDE_r	FED_r	π_{rr}
1	1.180	0.790	0.534	0.839	0.270	0.885	0.006	0.387
2	0.959	0.850	0.379	0.879	0.403	0.909	0.428	0.397
3	0.971	0.789	0.033	0.840	0.253	0.891	0.473	0.377
4	0.987	0.847	0.521	0.867	0.301	0.896	0.081	0.397
5	0.988	0.790	0.213	0.839	0.195	0.888	0.177	0.385
6	0.987	0.860	0.096	0.892	0.279	0.923	0.461	0.514
7	1.926	0.061	0.003	0.572	0.120	1.092	0.237	0.044
8	0.953	0.488	0.012	0.488	0.011	0.488	0.010	0.239
9	1.046	0.507	0.063	0.561	0.070	0.614	0.076	0.263
10	0.952	0.488	0.016	0.488	0.009	0.488	0.002	0.239
11	1.048	0.506	0.087	0.559	0.051	0.613	0.014	0.263
12	0.952	0.488	0.007	0.487	0.005	0.487	0.004	0.239
13	1.049	0.500	0.036	0.556	0.032	0.612	0.031	0.263

TABLE 92(S).—Pyridoxine.

Atom	P_π	SDN_r	FOD_r	SDR_r	FRD_r	SDE_r	FED_r	π_{rr}
1	1.187	0.761	0.599	0.828	0.303	0.895	0.007	0.383
2	0.956	0.841	0.327	0.880	0.384	0.919	0.440	0.399
3	0.981	0.765	0.082	0.835	0.286	0.704	0.489	0.375
4	0.985	0.826	0.572	0.866	0.330	0.925	0.085	0.399
5	0.990	0.269	0.129	0.836	0.154	0.902	0.179	0.386
6	0.993	0.847	0.203	0.889	0.340	0.932	0.477	0.414
7	1.928	0.058	0.008	0.572	0.127	1.087	0.247	0.043
8	1.080	0.276	0.001	0.332	0.004	0.387	0.007	0.161
9	0.912	0.369	0.025	0.372	0.032	0.375	0.039	0.172
10	1.080	0.276	0.001	0.332	0.001	0.387	0.001	0.161
11	0.914	0.368	0.044	0.371	0.026	0.374	0.008	0.172
12	1.080	0.276	0.000	0.332	0.001	0.388	0.003	0.161
13	0.914	0.365	0.010	0.369	0.013	0.374	0.016	0.172

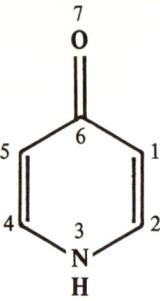

TABLE 93.—4-Pyridone.

Atom	P_{rr}	SDN_r	FOD_r	SDR_r	FRD_r	SDE_r	FED_r	π_{rr}
1	1.159	0.667	0.500	1.066	0.498	1.464	0.496	0.439
2	0.887	1.042	0.500	0.855	0.264	0.728	0.029	0.413
3	1.567	0.405	0.000	0.897	0.230	1.389	0.460	0.277
4	0.887	1.042	0.500	0.885	0.264	0.728	0.029	0.413
5	1.159	0.667	0.500	1.066	0.498	1.464	0.496	0.439
6	0.870	0.670	0.000	0.513	0.023	0.356	0.067	0.214
7	1.417	0.367	0.000	0.765	0.222	1.162	0.444	0.235

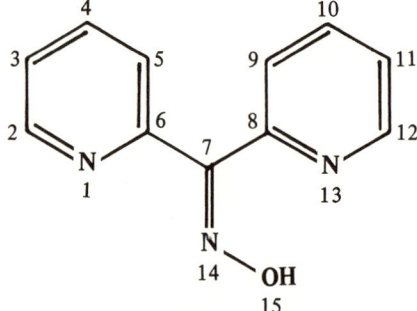

TABLE 94.—Di-2-Pyridylketoxime.

Atom	P_{rr}	SDN_r	FOD_r	SDR_r	FRD_r	SDE_r	FED_r	π_{rr}
1	1.163	0.949	0.136	0.926	0.105	0.904	0.075	0.410
2	0.935	0.946	0.000	0.845	0.020	0.745	0.039	0.397
3	1.001	0.940	0.133	0.918	0.142	0.896	0.152	0.414
4	0.962	0.944	0.031	0.843	0.016	0.742	0.001	0.396
5	0.996	0.959	0.080	0.936	0.109	0.914	0.138	0.428
6	0.951	0.883	0.094	0.782	0.078	0.682	0.062	0.343
7	0.972	0.994	0.262	0.904	0.255	0.815	0.248	0.370
8	0.951	0.883	0.094	0.782	0.078	0.682	0.062	0.343
9	0.996	0.959	0.080	0.936	0.109	0.914	0.138	0.428
10	0.962	0.944	0.031	0.843	0.016	0.742	0.001	0.396
11	1.001	0.940	0.133	0.918	0.142	0.896	0.152	0.414
12	0.935	0.946	0.000	0.845	0.020	0.745	0.039	0.397
13	1.163	0.949	0.136	0.926	0.105	0.904	0.075	0.410
14	1.094	1.680	0.717	1.477	0.617	1.273	0.025	0.627
15	1.917	0.167	0.074	0.636	0.135	1.104	0.196	0.066

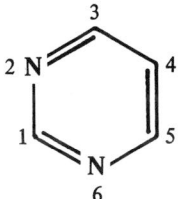

TABLE 95.—Pyrimidine.

Atom	P_{rr}	SDN_r	FOD_r	SDR_r	FRD_r	SDE_r	FED_r	π_{rr}
1	0.875	1.052	0.000	0.852	0.257	0.652	0.514	0.390
2	1.161	0.821	0.402	0.821	0.273	0.821	0.145	0.382
3	0.898	1.063	0.598	0.863	0.407	0.663	0.216	0.394
4	1.007	0.832	0.000	0.832	0.382	0.832	0.765	0.398
5	0.898	1.063	0.598	0.863	0.407	0.663	0.216	0.394
6	1.161	0.821	0.402	0.821	0.273	0.821	0.145	0.382

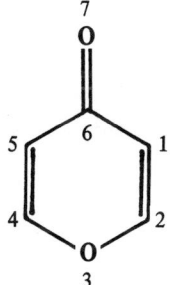

TABLE 96.—4-Pyrone.

Atom	P_{rr}	SDN_r	FOD_r	SDR_r	FRD_r	SDE_r	FED_r	π_{rr}
1	1.129	0.706	0.054	0.973	0.281	1.240	0.508	0.432
2	0.849	1.178	0.554	0.968	0.342	0.757	0.128	0.442
3	1.756	0.257	0.217	0.586	0.236	0.916	0.254	0.134
4	0.849	1.178	0.554	0.968	0.342	0.757	0.128	0.442
5	1.129	0.706	0.054	0.973	0.281	1.240	0.508	0.432
6	0.850	0.726	0.293	0.516	0.160	0.305	0.025	0.208
7	1.430	0.448	0.273	0.697	0.362	0.946	0.450	0.247

TABLE 97(P).—Quinon-Quinol X₃-POP.

Atom	P_{rr}	SDN_r	FOD_r	SDR_r	FRD_r	SDE_r	FED_r	π_{rr}
1	0.911	0.927	0.007	0.842	0.003	0.757	0.000	0.369
2	1.135	0.633	0.000	0.933	0.000	1.234	0.000	0.406
3	0.916	0.909	0.007	0.824	0.003	0.740	0.000	0.362
4	1.135	0.633	0.000	0.933	0.000	1.234	0.000	0.406
5	0.911	0.927	0.007	0.842	0.003	0.757	0.000	0.369
6	1.136	0.584	0.001	0.856	0.001	1.129	0.000	0.357
7	0.829	0.961	0.029	0.622	0.014	0.283	0.000	0.207
8	0.992	0.222	0.223	1.655	0.115	1.088	0.007	0.546
9	0.925	1.455	0.057	1.117	0.029	0.778	0.000	0.436
10	0.944	1.827	0.191	1.260	0.099	0.693	0.006	0.365
11	0.983	3.127	0.398	2.175	0.224	1.224	0.050	0.642
12	0.817	1.417	0.119	0.850	0.059	0.283	0.000	0.211
13	0.829	1.189	0.075	0.755	0.047	0.320	0.020	0.216
14	1.156	1.491	0.138	1.706	0.357	1.921	0.577	0.583
15	0.922	1.515	0.107	1.080	0.068	0.645	0.029	0.357
16	1.161	0.650	0.013	1.080	0.194	1.512	0.375	0.380
17	0.860	0.752	0.003	0.560	0.019	0.368	0.034	0.216
18	1.222	0.633	0.002	1.195	0.065	1.757	0.128	0.445
19	0.873	1.090	0.004	0.898	0.003	0.706	0.001	0.373
20	1.135	0.524	0.003	0.827	0.078	1.130	0.153	0.325
21	0.836	1.575	0.063	1.154	0.060	0.733	0.057	0.402
22	0.642	2.211	0.032	1.421	0.082	0.630	0.133	0.423
23	1.920	0.078	0.001	0.561	0.001	1.044	0.000	0.049
24	1.918	0.080	0.001	0.563	0.001	1.046	0.000	0.050
25	1.257	2.098	0.262	1.356	0.132	0.614	0.002	0.296

TABLE 97(P).—*Continued.*

26	1.303	1.538	0.164	1.164	0.146	0.789	0.127	0.302
27	1.918	0.080	0.001	0.563	0.001	1.046	0.000	0.052
28	1.348	0.967	0.063	0.858	0.032	0.750	0.001	0.275
29	1.433	0.535	0.008	0.835	0.112	1.134	0.217	0.260
30	1.897	0.191	0.011	0.606	0.015	1.021	0.018	0.069
31	1.909	0.105	0.001	0.566	0.001	1.027	0.000	0.057
32	1.857	0.279	0.006	0.619	0.024	0.959	0.043	0.099
33	0.956	0.485	0.000	0.485	0.000	0.485	0.001	0.239
34	1.014	0.665	0.004	0.618	0.011	0.571	0.019	0.270

TABLE 97(S).—Quinon-Quinol X_3-POP.

Atom	P_{rr}	SDN_r	FOD_r	SDR_r	FRD_r	SDE_r	FED_r	π_{rr}
1	0.911	0.927	0.007	0.842	0.003	0.757	0.000	0.369
2	1.135	0.633	0.000	0.933	0.000	1.234	0.000	0.406
3	0.916	0.909	0.007	0.824	0.003	0.740	0.000	0.362
4	1.135	0.633	0.000	0.933	0.000	1.234	0.000	0.406
5	0.911	0.927	0.007	0.842	0.003	0.757	0.000	0.369
6	1.136	0.584	0.001	0.856	0.001	1.129	0.000	0.359
7	0.829	0.961	0.029	0.622	0.015	0.283	0.000	0.207
8	0.992	0.220	0.225	1.654	0.116	1.088	0.007	0.546
9	0.925	1.435	0.058	1.116	0.029	0.778	0.000	0.436
10	0.944	1.825	0.192	1.259	0.009	0.693	0.006	0.365
11	0.983	3.122	0.399	2.127	0.225	1.223	0.005	0.642
12	0.817	1.415	0.119	0.849	0.060	0.283	0.000	0.211
13	0.829	1.186	0.074	0.753	0.047	0.320	0.020	0.216
14	1.157	1.475	0.136	1.701	0.357	1.927	0.577	0.582
15	0.922	1.511	0.107	1.078	0.068	0.645	0.029	0.357
16	1.162	0.645	0.014	1.082	0.195	1.518	0.377	0.379
17	0.862	0.748	0.003	0.558	0.019	0.369	0.034	0.216
18	1.224	0.624	0.002	1.194	0.066	1.764	0.131	0.442
19	0.872	1.084	0.005	0.895	0.003	0.705	0.001	0.373
20	1.141	0.510	0.004	0.825	0.080	1.141	0.156	0.322
21	0.836	1.555	0.062	1.143	0.059	0.732	0.057	0.401
22	0.628	2.159	0.029	1.390	0.081	0.621	0.132	0.416
23	1.920	0.078	0.001	0.561	0.001	1.044	0.000	0.049
24	1.918	0.080	0.001	0.563	0.001	1.046	0.000	0.050
25	1.257	2.093	0.262	1.353	0.132	0.614	0.002	0.296
26	1.304	1.529	0.164	1.160	0.146	0.791	0.128	0.302
27	1.918	0.080	0.001	0.563	0.001	1.046	0.000	0.050
28	1.348	0.966	0.064	0.858	0.032	0.750	0.001	0.275
29	1.435	0.537	0.008	0.834	0.114	1.141	0.219	0.259
30	1.897	0.187	0.011	0.604	0.014	1.021	0.018	0.069
31	1.909	0.103	0.001	0.565	0.001	1.027	0.000	0.057
32	1.856	0.267	0.005	0.611	0.024	0.955	0.042	0.099
33	1.080	0.275	0.000	0.329	0.000	0.383	0.001	0.160
34	0.891	0.459	0.002	0.404	0.006	0.350	0.010	0.173

TABLE 98(P).—*all-trans* Retinal.

Atom	P_π	SDN_r	FOD_r	SDR_r	FRD_r	SDE_r	FED_r	π_{rr}
1	1.039	0.647	0.041	0.661	0.039	0.675	0.036	0.272
2	0.953	0.484	0.001	0.484	0.001	0.484	0.001	0.238
3	0.905	2.061	0.336	1.770	0.307	1.479	0.277	0.631
4	0.953	0.484	0.001	0.484	0.001	0.484	0.001	0.238
5	1.039	0.647	0.041	0.661	0.039	0.675	0.036	0.272
6	1.032	0.834	0.013	0.931	0.036	1.028	0.058	0.409
7	0.950	1.785	0.306	1.494	0.258	1.203	0.209	0.508
8	1.029	1.045	0.060	1.143	0.099	1.240	0.139	0.457
9	0.915	1.591	0.245	1.299	0.184	1.008	0.123	0.447
10	1.046	1.120	0.107	1.309	0.189	1.497	0.271	0.481
11	0.945	1.463	0.175	1.171	0.108	0.880	0.041	0.444
12	1.057	1.275	0.181	1.464	0.258	1.653	0.334	0.511
13	0.890	1.266	0.101	0.975	0.051	0.604	0.001	0.386
14	1.128	1.403	0.229	1.677	0.291	1.951	0.253	0.590
15	0.813	0.921	0.042	0.630	0.030	0.338	0.018	0.224
16	1.322	0.779	0.080	0.840	0.093	0.828	0.085	0.266
17	0.953	0.487	0.000	0.487	0.000	0.487	0.000	0.239
18	1.040	0.595	0.030	0.610	0.023	0.624	0.016	0.266
19	0.953	0.488	0.000	0.488	0.000	0.488	0.000	0.240
20	1.038	0.557	0.012	0.572	0.006	0.586	0.000	0.264

TABLE 98(S).—*all-trans* Retinal.

Atom	P_{rr}	SDN_r	FOD_r	SDR_r	FRD_r	SDE_r	FED_r	π_{rr}
1	0.906	0.451	0.025	0.431	0.023	0.411	0.020	0.175
2	1.079	0.276	0.001	0.331	0.001	0.386	0.001	0.160
3	0.871	2.038	0.357	1.759	0.316	1.479	0.274	0.641
4	1.079	0.276	0.001	0.331	0.001	0.386	0.001	0.160
5	0.906	0.451	0.025	0.431	0.023	0.411	0.020	0.175
6	1.055	0.750	0.006	0.914	0.036	1.077	0.066	0.401
7	0.940	1.753	0.334	1.473	0.269	1.193	0.204	0.508
8	1.044	0.965	0.052	1.121	0.100	1.290	0.147	0.451
9	0.902	1.562	0.268	1.282	0.192	1.003	0.119	0.449
10	1.063	1.011	0.092	1.292	0.188	1.572	0.285	0.473
11	0.942	1.434	0.192	1.152	0.115	0.875	0.038	0.443
12	1.071	1.163	0.176	1.443	0.261	1.724	0.346	0.503
13	0.882	1.244	0.107	0.964	0.054	0.684	0.001	0.387
14	1.146	1.249	0.218	1.636	0.290	2.024	0.362	0.571
15	0.814	0.902	0.042	0.623	0.030	0.343	0.019	0.225
16	1.326	0.733	0.079	0.788	0.083	0.846	0.087	0.265
17	1.080	0.277	0.000	0.332	0.000	0.387	0.001	0.161
18	0.908	0.420	0.019	0.400	0.014	0.380	0.009	0.173
19	1.080	0.276	0.000	0.331	0.000	0.380	0.000	0.161
20	0.907	0.398	0.008	0.378	0.004	0.358	0.000	0.172

TABLE 99(P).—*all-trans* Retinal, Schiff-Base.

Atom	P_{rr}	SDN_r	FOD_r	SDR_r	FRD_r	SDE_r	FED_r	π_{rr}
1	1.038	0.663	0.038	0.669	0.035	0.676	0.032	0.272
2	0.953	0.484	0.000	0.484	0.000	0.484	0.001	0.238
3	0.900	2.197	0.308	1.840	0.278	1.482	0.248	0.637
4	0.953	0.484	0.000	0.484	0.000	0.484	0.001	0.238
5	1.038	0.663	0.038	0.669	0.035	0.676	0.032	0.272
6	1.032	0.832	0.008	0.028	0.027	1.025	0.047	0.409
7	0.946	1.925	0.289	1.568	0.241	1.211	0.193	0.513
8	1.028	1.037	0.040	1.134	0.076	1.230	0.112	0.455
9	0.910	1.740	0.246	1.382	0.184	1.025	0.123	0.453
10	1.043	1.099	0.072	1.285	0.148	1.471	0.226	0.476
11	0.943	1.631	0.194	1.273	0.122	0.916	0.050	0.454
12	1.065	1.118	0.121	1.392	0.209	1.667	0.296	0.487
13	0.894	1.477	0.134	1.120	0.070	0.763	0.006	0.411
14	1.093	1.274	0.165	1.542	0.243	1.810	0.321	0.518
15	0.871	1.319	0.080	0.962	0.042	0.605	0.005	0.379
16	1.324	1.536	0.216	1.804	0.258	2.072	0.300	0.639
17	0.953	0.487	0.000	0.487	0.000	0.487	0.000	0.239
18	1.040	0.613	0.030	0.620	0.023	0.626	0.016	0.267
19	0.953	0.487	0.000	0.487	0.000	0.487	0.000	0.239
20	1.038	0.582	0.016	0.588	0.009	0.595	0.001	0.265

TABLE 99(S).—*all-trans* Retinal, Schiff-Base.

Atom	P_{rr}	SDN_r	FOD_r	SDR_r	FRD_r	SDE_r	FED_r	π_{rr}
1	0.905	0.460	0.023	0.435	0.021	0.411	0.018	0.175
2	1.079	0.277	0.000	0.331	0.001	0.386	0.001	0.160
3	0.865	2.159	0.327	1.819	0.285	1.499	0.243	0.646
4	1.079	0.277	0.000	0.331	0.001	0.386	0.001	0.160
5	0.905	0.460	0.023	0.435	0.021	0.411	0.018	0.175
6	1.054	0.751	0.003	0.912	0.029	1.073	0.054	0.401
7	0.935	1.878	0.313	1.538	0.250	1.199	0.187	0.512
8	1.043	0.957	0.034	1.118	0.077	1.280	0.119	0.450
9	0.897	1.697	0.267	1.357	0.192	1.017	0.117	0.454
10	1.059	0.994	0.058	1.269	0.148	1.544	0.239	0.468
11	0.939	1.588	0.212	1.249	0.129	0.909	0.046	0.453
12	1.051	1.227	0.127	1.412	0.207	1.698	0.286	0.495
13	0.885	1.442	0.144	1.102	0.075	0.762	0.005	0.413
14	1.109	1.131	0.153	1.507	0.242	1.882	0.330	0.505
15	0.871	1.287	0.084	0.947	0.045	0.607	0.006	0.379
16	1.336	1.387	0.210	1.762	0.259	2.138	0.308	0.625
17	1.080	0.277	0.000	0.332	0.000	0.287	0.001	0.161
18	0.908	0.429	0.019	0.405	0.014	0.381	0.009	0.173
19	1.080	0.276	0.000	0.331	0.000	0.386	0.000	0.161
20	0.907	0.411	0.010	0.387	0.005	0.363	0.000	0.173

TABLE 100(P).—*all-trans* Retinol.

Atom	P_π	SDN$_r$	FOD$_r$	SDR$_r$	FRD$_r$	SDE$_r$	FED$_r$	π_π
1	1.044	0.600	0.043	0.662	0.040	0.725	0.037	0.272
2	0.953	0.484	0.001	0.484	0.001	0.484	0.000	0.238
3	0.906	1.673	0.343	1.776	0.316	1.879	0.289	0.639
4	0.953	0.484	0.001	0.484	0.001	0.484	0.000	0.238
5	1.044	0.600	0.043	0.662	0.040	0.725	0.037	0.272
6	1.031	0.840	0.027	0.941	0.034	1.042	0.041	0.410
7	0.995	1.388	0.286	1.491	0.265	1.594	0.243	0.514
8	1.024	1.061	0.109	1.162	0.102	1.263	0.096	0.459
9	0.965	1.183	0.188	1.286	0.184	1.389	0.181	0.453
10	1.033	1.169	0.190	1.373	0.200	1.578	0.209	0.491
11	1.009	1.029	0.090	1.133	0.098	1.236	0.106	0.446
12	1.021	1.356	0.279	1.560	0.279	1.764	0.280	0.535
13	0.994	0.789	0.019	0.892	0.031	0.995	0.044	0.380
14	0.986	1.624	0.315	1.933	0.338	2.423	0.361	0.716
15	0.953	0.483	0.001	0.483	0.001	0.483	0.011	0.238
16	1.048	0.591	0.039	0.679	0.043	0.767	0.046	0.273
17	0.953	0.487	0.001	0.487	0.000	0.487	0.000	0.239
18	1.046	0.546	0.023	0.609	0.023	0.671	0.023	0.266
19	0.952	0.488	0.000	0.488	0.000	0.488	0.000	0.240
20	1.049	0.501	0.002	0.563	0.004	0.626	0.006	0.263

TABLE 100(S).—*all-trans* Retinol.

Atom	P_{π}	SDN_r	FOD_r	SDR_r	FRD_r	SDE_r	FED_r	π_{π}
1	0.909	0.425	0.026	0.435	0.024	0.446	0.021	0.175
2	1.079	0.276	0.000	0.332	0.000	0.388	0.001	0.160
3	0.916	1.669	0.371	1.813	0.329	1.957	0.288	0.655
4	1.079	0.276	0.000	0.332	0.000	0.388	0.001	0.160
5	0.909	0.425	0.026	0.435	0.024	0.446	0.021	0.175
6	1.054	0.757	0.017	0.933	0.033	1.108	0.048	0.412
7	0.989	1.369	0.318	1.513	0.280	1.657	0.242	0.518
8	1.039	0.985	0.105	1.160	0.103	1.335	0.102	0.454
9	0.968	1.163	0.206	1.307	0.194	1.452	0.182	0.457
10	1.050	1.069	0.183	1.389	0.202	1.709	0.222	0.485
11	1.013	1.001	0.095	1.145	0.102	1.289	0.109	0.447
12	1.033	1.262	0.283	1.582	0.286	1.902	0.290	0.531
13	0.999	0.756	0.016	0.900	0.032	1.044	0.048	0.381
14	0.994	1.512	0.315	1.982	0.348	2.452	0.315	0.720
15	1.079	0.276	0.000	0.332	0.001	0.389	0.002	0.160
16	0.914	0.414	0.022	0.447	0.025	0.481	0.028	0.176
17	1.080	0.276	0.000	0.332	0.000	0.388	0.001	0.161
18	0.912	0.392	0.014	0.402	0.014	0.412	0.013	0.173
19	1.080	0.276	0.000	0.332	0.000	0.388	0.000	0.161
20	0.905	0.364	0.001	0.374	0.002	0.384	0.004	0.172

TABLE 101(P).—Riboflavin.

Atom	P_π	SDN_r	FOD_r	SDR_r	FRD_r	SDE_r	FED_r	π_π
1	0.992	0.812	0.027	0.889	0.101	0.966	0.175	0.391
2	0.940	1.098	0.157	0.955	0.099	0.812	0.040	0.399
3	1.059	0.878	0.074	0.981	0.077	1.085	0.081	0.432
4	0.920	1.009	0.093	0.866	0.105	0.723	0.118	0.361
5	1.565	0.739	0.188	1.064	0.243	1.389	0.297	0.325
6	0.832	1.151	0.115	0.819	0.058	0.488	0.002	0.325
7	1.462	0.499	0.067	1.232	0.400	1.965	0.733	0.356
8	0.876	0.717	0.022	0.507	0.019	0.297	0.016	0.205
9	1.754	0.163	0.003	0.832	0.006	1.500	0.008	0.151
10	0.814	0.823	0.055	0.550	0.030	0.278	0.006	0.205
11	0.976	1.279	0.287	1.084	0.218	0.890	0.149	0.411
12	1.033	1.950	0.564	1.334	0.285	0.718	0.006	0.455
13	1.006	0.679	0.000	0.756	0.072	0.833	0.144	0.336
14	1.000	1.182	0.194	1.067	0.100	0.952	0.006	0.453
15	1.380	0.610	0.093	0.687	0.071	0.764	0.050	0.249
16	1.443	0.419	0.038	0.669	0.086	0.919	0.135	0.228
17	0.952	0.488	0.000	0.488	0.001	0.488	0.002	0.239
18	1.049	0.503	0.003	0.562	0.015	0.622	0.026	0.263
19	0.953	0.488	0.000	0.488	0.000	0.488	0.000	0.239
20	1.043	0.537	0.020	0.569	0.013	0.602	0.006	0.264

TABLE 101(S).—Riboflavin.

Atom	P_{rr}	SDN_r	FOD_r	SDR_r	FRD_r	SDE_r	FED_r	π_{rr}
1	0.989	0.811	0.030	0.894	0.104	0.976	0.178	0.393
2	0.937	1.094	0.159	0.956	0.100	0.819	0.041	0.401
3	1.064	0.855	0.069	0.972	0.075	1.089	0.081	0.428
4	0.922	1.003	0.096	0.866	0.108	0.728	0.119	0.362
5	1.565	0.734	0.190	1.062	0.244	1.391	0.298	0.325
6	0.833	1.146	0.115	0.817	0.059	0.488	0.002	0.325
7	1.463	0.493	0.066	1.232	0.401	1.970	0.737	0.355
8	0.826	0.716	0.022	0.507	0.020	0.298	0.017	0.205
9	1.754	0.163	0.003	0.831	0.006	1.500	0.008	0.150
10	0.814	0.821	0.055	0.550	0.031	0.278	0.006	0.205
11	0.977	1.263	0.286	1.078	0.218	0.893	0.151	0.410
12	1.033	1.936	0.571	1.327	0.288	0.718	0.005	0.454
13	1.008	0.674	0.001	0.756	0.073	0.838	0.145	0.336
14	1.006	1.155	0.192	1.057	0.099	0.958	0.007	0.451
15	1.381	0.606	0.093	0.685	0.071	0.765	0.050	0.249
16	1.443	0.417	0.038	0.669	0.087	0.921	0.136	0.228
17	1.080	0.276	0.000	0.332	0.001	0.388	0.002	0.161
18	0.914	0.368	0.002	0.373	0.008	0.379	0.014	0.172
19	1.080	0.276	0.000	0.332	0.000	0.387	0.000	0.161
20	0.911	0.387	0.011	0.377	0.007	0.378	0.003	0.172

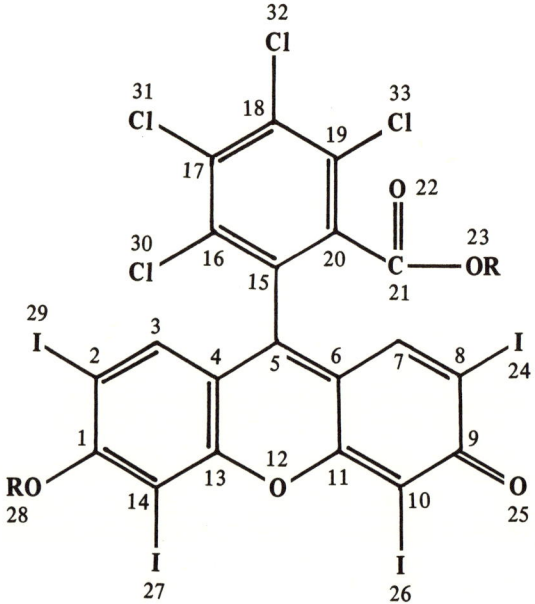

TABLE 102.—Rose Bengal.

Atom	P_{π}	SDN_r	FOD_r	SDR_r	FRD_r	SDE_r	FED_r	π_{π}
1	0.883	1.176	0.108	0.954	0.079	0.732	0.050	0.373
2	1.102	0.768	0.017	0.895	0.026	1.023	0.035	0.401
3	0.946	1.311	0.136	1.077	0.079	0.843	0.032	0.439
4	1.048	0.644	0.001	0.772	0.032	0.900	0.063	0.340
5	0.848	2.230	0.460	1.427	0.264	0.624	0.067	0.395
6	1.016	0.836	0.082	0.975	0.178	1.115	0.275	0.367
7	0.939	1.247	0.114	1.013	0.057	0.779	0.000	0.422
8	1.137	1.117	0.137	1.256	0.203	1.395	0.268	0.490
9	0.869	0.779	0.050	0.583	0.050	0.397	0.050	0.217
10	1.263	0.634	0.022	1.236	0.245	1.837	0.468	0.434
11	0.883	1.108	0.073	0.890	0.038	0.672	0.003	0.366
12	1.804	0.332	0.052	0.656	0.038	0.990	0.023	0.119
13	0.884	1.168	0.090	0.950	0.068	0.732	0.046	0.379
14	1.139	0.708	0.009	0.900	0.005	1.091	0.000	0.402
15	1.008	0.792	0.024	0.814	0.026	0.836	0.027	0.354
16	1.007	1.191	0.121	1.061	0.081	0.931	0.042	0.421
17	1.030	0.868	0.013	0.883	0.007	0.897	0.000	0.391
18	1.017	1.154	0.130	1.036	0.089	0.918	0.048	0.410
19	1.013	0.890	0.005	0.888	0.012	0.887	0.018	0.394
20	1.028	1.137	0.136	1.039	0.076	0.490	0.015	0.404
21	0.790	0.843	0.079	0.552	0.010	0.262	0.000	0.202
22	1.353	0.603	0.036	0.627	0.020	0.656	0.003	0.243
23	1.924	0.065	0.003	0.506	0.002	0.947	0.000	0.042
24	1.986	0.029	0.005	0.806	0.018	1.582	0.031	0.014
25	1.450	0.644	0.095	0.933	0.214	1.223	0.332	0.266
26	1.990	0.011	0.001	0.813	0.028	1.614	0.055	0.008
27	1.988	0.011	0.000	0.791	0.000	1.570	0.000	0.010
28	1.912	0.125	0.017	0.580	0.017	1.035	0.016	0.056
29	1.987	0.013	0.001	0.789	0.002	1.565	0.004	0.010
30	1.923	0.126	0.018	0.656	0.016	1.186	0.014	0.056
31	1.931	0.071	0.002	0.629	0.001	1.187	0.000	0.046
32	1.926	0.122	0.020	0.654	0.018	1.186	0.016	0.053
33	1.929	0.071	0.001	0.626	0.003	1.182	0.006	0.048

TABLE 103.—Rutin.

Atom	P_{rr}	SDN_r	FOD_r	SDR_r	FRD_r	SDE_r	FED_r	π_{rr}
1	1.131	0.652	0.006	0.936	0.014	1.230	0.023	0.407
2	0.921	0.860	0.060	0.806	0.034	0.752	0.008	0.361
3	1.134	0.645	0.026	0.963	0.039	1.281	0.052	0.410
4	0.923	0.855	0.020	0.800	0.010	0.746	0.000	0.364
5	1.823	0.179	0.062	0.630	0.070	1.081	0.079	0.104
6	0.968	0.934	0.354	1.041	0.306	1.148	0.257	0.420
7	1.062	0.830	0.264	1.298	0.417	1.767	0.570	0.472
8	0.848	0.776	0.173	0.560	0.098	0.343	0.024	0.214
9	1.125	0.588	0.001	0.840	0.001	1.091	0.001	0.356
10	0.915	0.891	0.086	0.837	0.054	0.783	0.021	0.371
11	1.034	0.684	0.121	0.791	0.109	0.897	0.096	0.339
12	1.052	0.857	0.184	0.987	0.123	1.117	0.062	0.442
13	0.986	0.747	0.005	0.853	0.026	0.960	0.047	0.375
14	0.891	0.830	0.205	0.964	0.199	1.099	0.193	0.395
15	1.048	0.756	0.008	0.863	0.018	0.970	0.028	0.304
16	1.029	0.866	0.155	1.001	0.148	1.136	0.141	0.437
17	1.921	0.069	0.007	0.558	0.005	1.047	0.002	0.048
18	1.929	0.056	0.001	0.578	0.007	1.099	0.014	0.042
19	1.926	0.070	0.024	0.597	0.041	1.124	0.059	0.045
20	1.918	0.073	0.010	0.562	0.008	1.051	0.000	0.049
21	1.399	0.538	0.201	0.769	0.171	1.000	0.141	0.263
22	1.929	0.075	0.030	0.670	0.103	1.265	0.175	0.047

TABLE 104(P).—Serotonin.

Atom	P_π	SDN_r	FOD_r	SDR_r	FRD_r	SDE_r	FED_r	π_π
1	0.982	0.725	0.040	0.848	0.069	0.971	0.098	0.368
2	1.068	0.748	0.302	0.928	0.200	1.107	0.098	0.414
3	1.034	0.828	0.457	0.951	0.334	1.074	0.212	0.424
4	1.041	0.641	0.001	0.820	0.006	0.999	0.010	0.357
5	1.628	0.308	0.086	0.915	0.213	1.522	0.341	0.237
6	1.021	0.862	0.417	1.040	0.307	1.219	0.196	0.461
7	1.132	0.556	0.089	1.047	0.313	1.359	0.538	0.383
8	1.064	0.611	0.144	0.734	0.072	0.857	0.000	0.331
9	1.085	0.784	0.443	1.063	0.418	1.342	0.392	0.447
10	1.930	0.052	0.004	0.577	0.019	1.102	0.034	0.040
11	0.952	0.487	0.003	0.487	0.004	0.487	0.004	0.239
12	1.064	0.471	0.015	0.582	0.046	0.692	0.078	0.263

TABLE 104(S).—Serotonin.

Atom	P_π	SDN_r	FOD_r	SDR_r	FRD_r	SDE_r	FED_r	π_π
1	0.982	0.724	0.038	0.848	0.068	0.974	0.099	0.368
2	1.068	0.749	0.303	0.929	0.207	1.110	0.100	0.415
3	1.034	0.828	0.466	0.953	0.339	1.078	0.212	0.424
4	1.041	0.640	0.001	0.821	0.006	1.002	0.010	0.357
5	1.631	0.302	0.081	0.926	0.214	1.538	0.347	0.235
6	1.031	0.838	0.396	1.040	0.300	1.242	0.204	0.458
7	1.128	0.563	0.086	1.064	0.317	1.560	0.549	0.388
8	1.066	0.607	0.155	0.732	0.078	0.857	0.000	0.330
9	1.085	0.785	0.454	1.067	0.425	1.349	0.396	0.448
10	1.930	0.052	0.004	0.577	0.019	1.103	0.034	0.040
11	1.080	0.276	0.000	0.333	0.002	0.390	0.004	0.161
12	0.923	0.350	0.007	0.386	0.025	0.422	0.043	0.173

TABLE 105(P).—Skatole.

Atom	P_{rr}	SDN_r	FOD_r	SDR_r	FRD_r	SDE_r	FED_r	π_{rr}
1	1.065	0.610	0.131	0.727	0.067	0.844	0.003	0.331
2	1.021	0.873	0.505	0.990	0.399	1.107	0.294	0.437
3	1.040	0.745	0.063	0.862	0.034	0.979	0.046	0.399
4	1.031	0.785	0.245	0.902	0.219	1.020	0.193	0.407
5	1.038	0.826	0.458	0.932	0.319	1.060	0.181	0.424
6	1.019	0.678	0.008	0.796	0.031	0.912	0.054	0.365
7	1.624	0.313	0.070	0.891	0.181	1.624	0.292	0.239
8	1.020	0.864	0.405	1.039	0.349	1.214	0.292	0.460
9	1.127	0.564	0.096	1.033	0.329	1.501	0.561	0.384
10	0.952	0.487	0.005	0.487	0.004	0.487	0.005	0.239
11	1.064	0.472	0.016	0.580	0.049	0.687	0.082	0.263

TABLE 105(S).—Skatole.

Atom	P_{rr}	SDN_r	FOD_r	SDR_r	FRD_r	SDE_r	FED_r	π_{rr}
1	1.067	0.606	0.142	0.725	0.072	0.845	0.003	0.330
2	1.021	0.874	0.516	0.993	0.407	1.112	0.298	0.438
3	1.041	0.743	0.060	0.862	0.053	0.981	0.046	0.399
4	1.031	0.785	0.256	0.905	0.226	1.024	0.196	0.418
5	1.038	0.825	0.467	0.944	0.325	1.063	0.183	0.424
6	1.020	0.677	0.007	0.796	0.037	0.915	0.055	0.356
7	1.627	0.307	0.066	0.896	0.182	1.482	0.298	0.237
8	1.030	0.840	0.384	1.039	0.342	1.238	0.300	0.458
9	1.122	0.572	0.093	1.048	0.333	1.525	0.572	0.389
10	1.080	0.276	0.000	0.333	0.002	0.389	0.005	0.161
11	0.923	0.351	0.007	0.385	0.026	0.419	0.045	0.173

TABLE 106.—Sphandin.

Atom	P_{rr}	SDN_r	FOD_r	SDR_r	FRD_r	SDE_r	FED_r	π_{rr}
1	1.114	0.653	0.009	0.946	0.114	1.240	0.220	0.401
2	0.977	0.951	0.013	1.090	0.105	1.228	0.198	0.486
3	1.786	0.214	0.035	0.577	0.062	0.939	0.089	0.119
4	1.007	0.807	0.185	0.907	0.095	1.006	0.005	0.394
5	0.995	0.740	0.006	0.913	0.146	1.086	0.286	0.281
6	1.074	0.851	0.207	1.063	0.204	1.274	0.201	0.455
7	1.060	0.603	0.043	0.776	0.077	0.949	0.111	0.332
8	0.914	1.205	0.551	0.974	0.276	0.742	0.002	0.424
9	1.082	0.966	0.450	1.131	0.285	1.311	0.125	0.491
10	0.814	0.763	0.104	0.531	0.059	0.300	0.013	0.208
11	1.828	0.177	0.064	0.582	0.080	0.987	0.095	0.097
12	0.948	0.963	0.174	1.066	0.294	1.169	0.415	0.431
13	1.069	0.605	0.031	0.743	0.043	0.881	0.055	0.328
14	1.930	0.056	0.001	0.591	0.048	1.126	0.096	0.041
15	1.401	0.471	0.146	0.162	0.108	0.853	0.091	0.240

TABLE 107(P).—Tetrahydrobiopterin.

Atom	P_{rr}	SDN_r	FOD_r	SDR_r	FRD_r	SDE_r	FED_r	π_{rr}
1	0.774	1.080	0.263	0.949	0.157	0.817	0.050	0.321
2	1.438	0.371	0.012	1.646	0.099	2.922	0.187	0.323
3	0.903	0.951	0.405	1.253	0.264	1.556	0.123	0.374
4	1.205	0.490	0.003	3.255	0.313	6.021	0.623	0.413
5	1.175	0.822	0.357	0.944	0.194	1.065	0.030	0.389
6	1.061	0.951	0.547	5.124	0.523	5.296	0.545	0.626
7	1.787	0.207	0.072	1.101	0.062	1.994	0.051	0.172
8	1.787	0.207	0.072	1.101	0.062	1.994	0.051	0.172
9	1.816	0.194	0.110	1.439	0.118	2.685	0.125	0.162
10	0.953	0.485	0.000	0.485	0.000	0.485	0.000	0.239
11	1.075	0.460	0.000	0.848	0.039	1.237	0.078	0.266
12	0.954	0.484	0.010	0.484	0.005	0.484	0.000	0.238
13	1.059	0.511	0.079	0.828	0.073	1.144	0.068	0.271
14	0.954	0.484	0.010	0.484	0.005	0.484	0.000	0.238
15	1.059	0.511	0.079	0.828	0.073	1.144	0.068	0.271

TABLE 107(S).—Tetrahydrobiopterin.

Atom	P_{rr}	SDN_r	FOD_r	SDR_r	FRD_r	SDE_r	FED_r	π_{rr}
1	0.780	1.062	0.304	1.051	0.180	1.040	0.056	0.326
2	1.444	0.362	0.006	1.953	0.099	3.545	0.193	0.320
3	0.924	0.898	0.411	1.527	0.277	2.157	0.142	0.380
4	1.211	0.489	0.016	4.306	0.331	8.122	0.645	0.423
5	1.197	0.745	0.307	1.003	0.173	1.261	0.039	0.379
6	1.035	0.993	0.602	3.962	0.576	6.913	0.549	0.660
7	1.790	0.200	0.078	1.191	0.066	2.182	0.054	0.169
8	1.790	0.200	0.078	1.191	0.066	2.182	0.054	0.169
9	1.826	0.174	0.106	1.684	0.121	3.194	0.137	0.152
10	1.080	0.276	0.000	0.339	0.001	0.403	0.002	0.161
11	0.930	0.344	0.001	0.616	0.024	0.887	0.046	0.174
12	1.079	0.275	0.000	0.337	0.001	0.398	0.001	0.160
13	0.918	0.378	0.045	0.589	0.042	0.800	0.039	0.176
14	1.079	0.275	0.000	0.337	0.001	0.398	0.001	0.160
15	0.918	0.378	0.045	0.589	0.042	0.800	0.039	0.176

TABLE 108(P).—Thiamine.

Atom	P_{rr}	SDN_r	FOD_r	SDR_r	FRD_r	SDE_r	FED_r	π_{rr}
1	0.911	0.909	0.011	0.846	0.051	0.783	0.091	0.380
2	1.254	0.681	0.006	0.892	0.040	1.103	0.073	0.388
3	0.905	0.836	0.000	0.767	0.058	0.697	0.115	0.340
4	1.042	0.733	0.006	0.954	0.132	1.175	0.257	0.405
5	0.977	0.906	0.010	0.844	0.006	0.781	0.002	0.401
6	1.075	0.713	0.000	0.933	0.118	1.154	0.236	0.410
7	1.660	0.343	0.187	0.589	0.108	0.836	0.028	0.164
8	1.165	1.104	0.796	1.393	0.532	1.683	0.268	0.627
9	1.092	1.112	0.691	1.398	0.464	1.683	0.237	0.659
10	1.089	0.687	0.019	0.976	0.013	1.266	0.006	0.431
11	0.956	0.855	0.157	1.000	0.139	1.145	0.122	0.434
12	1.836	0.136	0.001	1.080	0.219	2.024	0.438	0.131
13	0.953	0.488	0.000	0.488	0.001	0.488	0.001	0.240
14	1.042	0.514	0.002	0.557	0.008	0.599	0.014	0.263
15	0.937	0.472	0.010	0.472	0.008	0.472	0.006	0.225
16	1.096	0.503	0.074	0.642	0.079	0.780	0.085	0.279
17	0.954	0.487	0.003	0.487	0.002	0.487	0.001	0.239
18	1.048	0.506	0.024	0.574	0.021	0.641	0.018	0.265
19	0.952	0.487	0.000	0.487	0.000	0.487	0.000	0.239
20	1.059	0.485	0.003	0.571	0.002	0.656	0.001	0.264

TABLE 108(S).—Thiamine.

Atom	P_{rr}	SDN_r	FOD_r	SDR_r	FRD_r	SDE_r	FED_r	π_{rr}
1	0.907	0.898	0.000	0.846	0.050	0.974	0.099	0.382
2	1.258	0.665	0.000	0.884	0.040	1.103	0.079	0.384
3	0.910	0.828	0.000	0.767	0.036	0.707	0.125	0.341
4	1.039	0.727	0.000	0.960	0.141	1.193	0.281	0.408
5	0.982	0.890	0.000	0.838	0.001	0.786	0.002	0.400
6	1.081	0.700	0.000	0.933	0.129	1.166	0.258	0.408
7	1.673	0.342	0.211	0.596	0.118	0.851	0.025	0.164
8	1.168	1.073	0.827	1.388	0.538	1.702	0.249	0.621
9	1.102	1.176	0.739	1.388	0.477	1.699	0.216	0.652
10	1.095	0.672	0.039	0.986	0.023	1.301	0.007	0.433
11	0.957	0.928	0.144	0.999	0.129	1.168	0.118	0.432
12	1.837	0.134	0.000	1.085	0.238	2.036	0.476	0.130
13	1.080	0.276	0.000	0.331	0.000	0.387	0.001	0.161
14	0.909	0.373	0.000	0.370	0.004	0.366	0.008	0.172
15	1.068	0.272	0.000	0.329	0.003	0.386	0.005	0.156
16	0.941	0.373	0.026	0.419	0.034	0.464	0.043	0.182
17	1.080	0.276	0.000	0.332	0.001	0.388	0.001	0.161
18	0.913	0.368	0.011	0.380	0.010	0.392	0.009	0.173
19	1.079	0.276	0.000	0.332	0.000	0.388	0.000	0.161
20	0.921	0.357	0.003	0.379	0.002	0.402	0.001	0.173

TABLE 109(P).—Thiochrome.

Atom	$P_{\pi\pi}$	SDN_r	FOD_r	SDR_r	FRD_r	SDE_r	FED_r	π_{rr}
1	1.067	0.680	0.020	3.742	0.149	6.804	0.277	0.449
2	1.117	0.652	0.006	1.704	0.037	2.776	0.067	0.443
3	1.445	0.508	0.123	8.158	0.392	15.808	0.661	0.650
4	0.962	0.857	0.174	1.909	0.138	2.960	0.102	0.423
5	1.458	0.358	0.004	5.279	0.215	10.200	0.426	0.334
6	0.925	0.831	0.241	0.807	0.124	0.783	0.008	0.330
7	1.274	0.667	0.402	1.926	0.252	3.174	0.102	0.406
8	0.851	0.997	0.090	0.828	0.046	0.659	0.001	0.366
9	1.255	0.656	0.139	1.916	0.122	3.177	0.104	0.393
10	0.933	0.991	0.587	1.065	0.303	1.139	0.019	0.409
11	1.044	0.738	0.077	1.999	0.093	3.260	0.108	0.419
12	0.943	0.468	0.009	0.468	0.005	0.468	0.000	0.225
13	1.548	0.310	0.051	1.254	0.045	2.197	0.040	0.236
14	0.953	0.486	0.000	0.486	0.000	0.486	0.000	0.239
15	1.060	0.484	0.003	0.909	0.019	1.334	0.034	0.266
16	0.952	0.486	0.000	0.486	0.000	0.486	0.000	0.239
17	1.063	0.480	0.001	0.659	0.005	0.838	0.008	0.265
18	0.954	0.488	0.003	0.488	0.001	0.488	0.000	0.240
19	1.030	0.525	0.015	0.554	0.007	0.584	0.000	0.263
20	1.125	0.452	0.053	1.104	0.047	1.757	0.042	0.281

TABLE 109(S).—Thiochrome.

Atom	$P_{\pi\pi}$	SDN_r	FOD_r	SDR_r	FRD_r	SDE_r	FED_r	π_{rr}
1	1.007	0.664	0.018	4.784	0.149	8.905	0.280	0.440
2	1.129	0.630	0.020	2.141	0.050	3.651	0.080	0.444
3	1.450	0.496	0.171	10.298	0.406	20.100	0.642	0.552
4	0.968	0.843	0.208	2.353	0.158	3.864	0.108	0.423
5	1.468	0.347	0.010	6.989	0.224	12.630	0.437	0.330
6	0.927	0.829	0.343	0.831	0.175	0.834	0.007	0.330
7	1.281	0.651	0.333	2.326	0.219	4.002	0.105	0.401
8	0.847	0.985	0.021	0.833	0.011	0.681	0.001	0.367
9	1.262	0.637	0.213	2.349	0.161	4.061	0.108	0.390
10	0.943	0.963	0.533	1.183	0.278	1.403	0.023	0.410
11	1.041	0.736	0.030	2.448	0.071	4.159	0.112	0.423
12	1.011	0.272	0.000	0.339	0.001	0.406	0.007	0.156
13	1.603	0.306	0.077	1.499	0.061	2.692	0.045	0.236
14	1.080	0.276	0.000	0.340	0.000	0.403	0.001	0.161
15	0.920	0.357	0.001	0.650	0.011	0.943	0.020	0.173
16	1.079	0.276	0.000	0.335	0.000	0.393	0.000	0.161
17	0.923	0.354	0.002	0.461	0.004	0.569	0.006	0.174
18	1.080	0.276	0.000	0.331	0.000	0.386	0.000	0.161
19	0.906	0.379	0.002	0.368	0.001	0.358	0.000	0.172
20	0.954	0.351	0.018	0.763	0.021	1.174	0.024	0.184

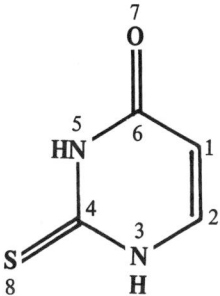

TABLE 110.—Thiouracil.

Atom	P_{rr}	SDN_r	FOD_r	SDR_r	FRD_r	SDE_r	FED_r	π_{rr}
1	1.218	0.587	0.339	1.259	0.381	1.931	0.423	0.432
2	0.856	1.129	0.988	0.919	0.511	0.708	0.035	0.411
3	1.669	0.272	0.150	1.102	0.285	1.931	0.419	0.225
4	0.925	0.703	0.025	0.703	0.050	0.703	0.075	0.313
5	1.747	0.175	0.079	1.005	0.143	1.834	0.207	0.163
6	0.829	0.714	0.201	0.504	0.101	0.293	0.002	0.206
7	1.432	0.395	0.176	0.644	0.096	0.893	0.014	0.232
8	1.324	0.461	0.041	1.386	0.434	2.711	0.826	0.425

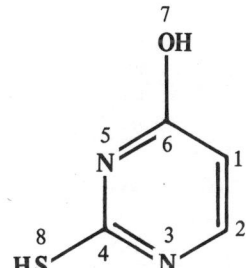

TABLE 111.—Thiouracil (Lactim form).

Atom	P_{rr}	SDN_r	FOD_r	SDR_r	FRD_r	SDE_r	FED_r	π_{rr}
1	1.122	0.654	0.046	1.997	0.093	3.341	0.141	0.417
2	0.900	1.027	0.738	0.842	0.369	0.657	0.000	0.389
3	1.269	0.638	0.308	1.982	0.224	3.326	0.140	0.398
4	0.915	0.865	0.020	0.865	0.017	0.865	0.014	0.358
5	1.292	0.619	0.390	1.963	0.263	3.307	0.137	0.394
6	0.852	0.968	0.446	0.783	0.223	0.598	0.000	0.345
7	1.912	0.081	0.043	0.543	0.022	1.006	0.000	0.053
8	1.738	0.228	0.009	14.117	0.788	28.006	1.568	0.359

7
O

9   10
C≡H₃

5   6   1
HN

4   3   2

O   N
      H
8

TABLE 112(P).—Thymine.

Atom	P_{rr}	SDN_r	FOD_r	SDR_r	FRD_r	SDE_r	FED_r	π_{rr}
1	1.164	0.584	0.324	1.150	0.568	1.714	0.812	0.414
2	0.880	1.113	0.979	0.976	0.609	0.838	0.239	0.435
3	1.683	0.256	0.154	0.952	0.334	1.648	0.513	0.209
4	0.809	0.698	0.021	0.491	0.017	0.284	0.012	0.204
5	1.750	0.166	0.074	0.826	0.038	1.485	0.003	0.153
6	0.830	0.717	0.191	0.510	0.105	0.302	0.019	0.208
7	1.429	0.397	0.168	0.655	0.166	0.912	0.165	0.233
8	1.438	0.338	0.019	0.596	0.063	0.853	0.107	0.217
9	0.951	0.486	0.013	0.486	0.010	0.486	0.008	0.239
10	1.067	0.473	0.058	0.592	0.090	0.711	0.121	0.264

TABLE 112(S).—Thymine.

Atom	P_{rr}	SDN_r	FOD_r	SDR_r	FRD_r	SDE_r	FED_r	π_{rr}
1	1.592	0.596	0.339	1.172	0.587	1.748	0.835	0.422
2	0.891	1.081	0.963	0.971	0.606	0.862	0.249	0.436
3	1.687	0.248	0.151	0.959	0.339	1.670	0.527	0.206
4	0.810	0.697	0.019	0.491	0.016	0.285	0.013	0.204
5	1.750	0.166	0.078	0.826	0.040	1.485	0.003	0.154
6	0.831	0.716	0.218	0.510	0.119	0.304	0.019	0.218
7	1.429	0.398	0.188	0.659	0.179	0.920	0.169	0.234
8	1.439	0.336	0.017	0.597	0.063	0.858	0.110	0.217
9	1.079	0.276	0.001	0.333	0.004	0.390	0.008	0.161
10	0.925	0.352	0.027	0.393	0.049	0.434	0.067	0.173

TABLE 113(P).—Thymine (Lactim form).

Atom	P_{rr}	SDN_r	FOD_r	SDR_r	FRD_r	SDE_r	FED_r	π_{rr}
1	1.049	0.686	0.063	0.877	0.362	1.068	0.662	0.388
2	0.914	1.023	0.754	0.865	0.428	0.712	0.102	0.401
3	1.245	0.659	0.274	0.850	0.301	1.040	0.328	0.377
4	0.839	0.949	0.037	0.786	0.155	0.623	0.273	0.344
5	1.262	0.650	0.411	0.837	0.218	1.025	0.026	0.374
6	0.861	0.970	0.406	0.807	0.311	0.644	0.217	0.355
7	1.912	0.082	0.039	0.548	0.076	1.015	0.113	0.054
8	1.911	0.076	0.004	0.543	0.073	1.010	0.141	0.053
9	0.952	0.448	0.002	0.488	0.010	0.488	0.017	0.239
10	1.055	0.488	0.011	0.561	0.066	0.635	0.120	0.263

TABLE 113(S).—Thymine (Lactim form).

Atom	P_{rr}	SDN_r	FOD_r	SDR_r	FRD_r	SDE_r	FED_r	π_{rr}
1	1.043	0.639	0.050	0.885	0.368	1.077	0.687	0.392
2	0.920	1.007	0.730	0.863	0.417	0.718	0.103	0.400
3	1.245	0.659	0.293	0.851	0.318	1.043	0.432	0.377
4	0.842	0.939	0.023	0.783	0.152	0.628	0.281	0.344
5	1.261	0.650	0.404	0.837	0.215	1.025	0.026	0.374
6	0.866	0.960	0.450	0.805	0.336	0.650	0.223	0.355
7	1.912	0.080	0.043	0.549	0.080	1.018	0.117	0.053
8	1.912	0.075	0.002	0.543	0.075	1.012	0.148	0.052
9	1.080	0.276	0.000	0.332	0.005	0.388	0.011	0.161
10	0.918	0.359	0.004	0.373	0.032	0.387	0.061	0.172

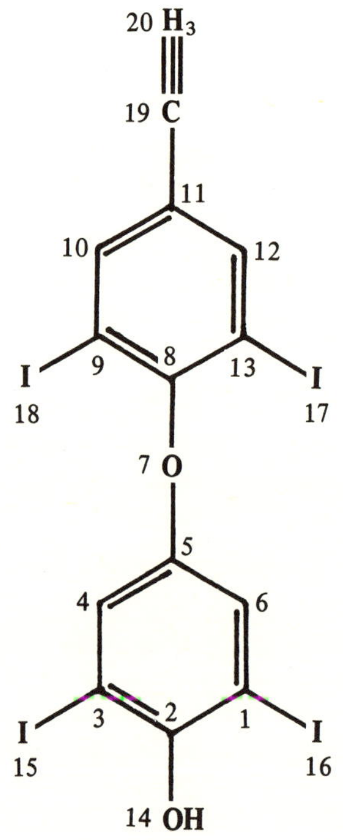
20 H3
19 C
11
10
12
9
8
13
I
18
I
17
7 O
5
4
6
3
2
1
I
15
I
16
14 OH

TABLE 114(P).—Thyroxine.

Atom	P_π	SDN_r	FOD_r	SDR_r	FRD_r	SDE_r	FED_r	π_π
1	1.087	0.762	0.001	0.867	0.046	0.971	0.090	0.400
2	0.954	0.776	0.000	0.854	0.145	0.933	0.291	0.372
3	1.087	0.762	0.001	0.867	0.046	0.971	0.090	0.400
4	1.026	0.817	0.002	0.897	0.070	0.977	0.139	0.415
5	0.985	0.734	0.000	0.839	0.128	0.943	0.256	0.369
6	1.026	0.817	0.002	0.897	0.070	0.977	0.139	0.415
7	1.865	0.104	0.000	0.625	0.134	1.164	0.268	0.078
8	0.939	0.817	0.000	0.816	0.053	0.618	0.107	0.367
9	1.086	0.767	0.475	0.873	0.290	0.978	0.105	0.402
10	0.995	0.861	0.514	0.861	0.268	0.861	0.022	0.408
11	0.999	0.748	0.000	0.854	0.082	0.960	0.164	0.383
12	0.995	0.861	0.514	0.861	0.268	0.861	0.022	0.408
13	1.086	0.767	0.475	0.873	0.290	0.978	0.105	0.402
14	1.926	0.058	0.000	0.574	0.056	1.090	0.112	0.043
15	1.987	0.012	0.000	0.787	0.007	1.561	0.015	0.011
16	1.987	0.012	0.000	0.787	0.007	1.561	0.015	0.011
17	1.987	0.012	0.008	0.787	0.013	1.562	0.017	0.011
18	1.987	0.012	0.008	0.787	0.013	1.562	0.017	0.011
19	0.952	0.488	0.000	0.488	0.001	0.488	0.002	0.240
20	1.050	0.496	0.000	0.558	0.013	0.621	0.025	0.263

TABLE 114(S).—Thyroxine.

Atom	P_π	SDN_r	FOD_r	SDR_r	FRD_r	SDE_r	FED_r	π_π
1	1.087	0.762	0.008	0.867	0.049	0.971	0.090	0.400
2	0.954	0.776	0.000	0.854	0.146	0.933	0.292	0.372
3	1.087	0.762	0.008	0.867	0.049	0.971	0.090	0.400
4	1.026	0.817	0.008	0.897	0.074	0.978	0.140	0.415
5	0.985	0.735	0.000	0.839	0.129	0.943	0.257	0.369
6	1.026	0.817	0.008	0.897	0.074	0.978	0.140	0.415
7	1.857	0.103	0.000	0.626	0.135	1.149	0.270	0.078
8	0.942	0.806	0.000	0.874	0.054	0.823	0.108	0.366
9	1.086	0.768	0.468	0.874	0.287	0.980	0.105	0.402
10	1.001	0.848	0.507	0.857	0.265	0.866	0.022	0.406
11	0.992	0.754	0.000	0.860	0.083	0.966	0.166	0.387
12	1.001	0.848	0.507	0.857	0.265	0.866	0.022	0.402
13	1.086	0.768	0.468	0.874	0.287	0.980	0.105	0.402
14	1.926	0.058	0.000	0.574	0.056	1.090	0.112	0.043
15	1.987	0.012	0.000	0.787	0.007	1.561	0.015	0.011
16	1.987	0.012	0.000	0.787	0.007	1.561	0.015	0.011
17	1.987	0.012	0.000	0.787	0.007	1.561	0.015	0.011
18	1.987	0.012	0.000	0.787	0.007	1.561	0.015	0.011
19	1.080	0.276	0.000	0.332	0.001	0.388	0.002	0.161
20	0.915	0.364	0.000	0.371	0.007	0.379	0.013	0.172

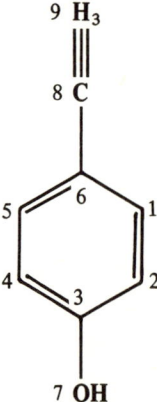

TABLE 115(P).—Tyrosine.

Atom	P_{rr}	SDN_r	FOD_r	SDR_r	FRD_r	SDE_r	FED_r	π_{rr}
1	1.012	0.825	0.500	0.850	0.307	0.875	0.115	0.404
2	1.049	0.769	0.500	0.870	0.361	0.973	0.221	0.408
3	0.952	0.789	0.000	0.814	0.234	0.840	0.469	0.366
4	1.049	0.769	0.500	0.870	0.361	0.973	0.221	0.408
5	1.012	0.825	0.500	0.850	0.307	0.875	0.115	0.404
6	0.999	0.744	0.000	0.847	0.259	0.949	0.517	0.383
7	1.925	0.060	0.000	0.565	0.118	1.070	0.237	0.044
8	0.952	0.488	0.000	0.488	0.006	0.488	0.012	0.240
9	1.050	0.495	0.000	0.558	.046	0.620	0.092	0.263

TABLE 115(S).—Tyrosine.

Atom	P_{rr}	SDN_r	FOD_r	SDR_r	FRD_r	SDE_r	FED_r	π_{rr}
1	1.017	0.811	0.500	0.845	0.308	0.880	0.116	0.402
2	1.049	0.769	0.500	0.872	0.364	0.975	0.228	0.408
3	0.956	0.778	0.000	0.812	0.240	0.847	0.480	0.366
4	1.049	0.769	0.500	0.872	0.364	0.975	0.228	0.408
5	1.017	0.811	0.500	0.845	0.308	0.880	0.116	0.402
6	0.993	0.750	0.000	0.852	0.266	0.955	0.523	0.386
7	1.926	0.059	0.000	0.566	0.122	1.072	0.244	0.043
8	1.080	0.276	0.000	0.332	0.004	0.388	0.008	0.161
9	0.915	0.363	0.000	0.371	0.023	0.378	0.047	0.172

11 O
13 HO
14 15 C≡≡H₃
16 17 C≡≡H₃
8 C
9 C
7 C
H
12 HO
2 1 6
10 O
20 H₂
18 19 C≡≡H₃
3 4 5

TABLE 116(P).—Ubiquinone.

Atom	P_{rr}	SDN_r	FOD_r	SDR_r	FRD_r	SDE_r	FED_r	π_{rr}
1	0.829	1.117	0.176	0.702	0.091	0.287	0.006	0.210
2	1.009	1.130	0.154	1.209	0.382	1.287	0.610	0.472
3	1.009	1.130	0.154	1.209	0.382	1.287	0.610	0.472
4	0.829	1.117	0.176	0.702	0.091	0.287	0.006	0.210
5	0.952	1.525	0.289	1.247	0.184	0.970	0.079	0.478
6	0.952	1.525	0.289	1.247	0.184	0.970	0.079	0.478
7	0.955	0.473	0.001	0.474	0.001	0.474	0.001	0.229
8	0.993	0.984	0.000	1.085	0.001	1.186	0.001	0.511
9	1.000	0.907	0.000	1.008	0.001	1.109	0.002	0.461
10	1.294	1.330	0.320	1.011	0.188	0.691	0.056	0.299
11	1.294	1.330	0.320	1.011	0.188	0.691	0.056	0.299
12	1.920	0.127	0.024	0.643	0.128	1.159	0.233	0.055
13	1.920	0.127	0.024	0.643	0.128	1.159	0.233	0.055
14	0.952	0.486	0.001	0.486	0.001	0.486	0.001	0.239
15	1.043	0.586	0.036	0.602	0.024	0.618	0.012	0.267
16	0.953	0.486	0.000	0.486	0.000	0.486	0.000	0.239
17	1.050	0.511	0.000	0.573	0.000	0.636	0.000	0.265
18	0.953	0.486	0.000	0.486	0.000	0.486	0.000	0.239
19	1.050	0.511	0.000	0.573	0.000	0.636	0.000	0.265
20	1.043	0.648	0.036	0.676	0.026	0.705	0.015	0.287

TABLE 116(S).—Ubiquinone.

Atom	P_{rr}	SDN_r	FOD_r	SDR_r	FRD_r	SDE_r	FED_r	π_{rr}
1	0.829	1.104	0.184	0.695	0.095	0.287	0.006	0.210
2	1.010	1.123	0.166	1.207	0.389	1.291	0.613	0.472
3	1.010	1.123	0.166	1.207	0.389	1.291	0.613	0.472
4	0.829	1.104	0.184	0.695	0.095	0.287	0.006	0.210
5	0.955	1.444	0.276	1.218	0.179	0.993	0.082	0.479
6	0.955	1.444	0.276	1.218	0.179	0.993	0.082	0.479
7	1.076	0.271	0.000	0.327	0.001	0.382	0.001	0.156
8	1.017	0.896	0.000	1.071	0.001	1.247	0.002	0.504
9	0.999	0.876	0.000	1.014	0.001	1.153	0.003	0.467
10	1.295	1.294	0.329	0.993	0.192	0.692	0.056	0.299
11	1.295	1.294	0.329	0.993	0.192	0.692	0.056	0.299
12	1.920	0.126	0.025	0.643	0.130	1.160	0.234	0.055
13	1.920	0.126	0.025	0.643	0.130	1.160	0.234	0.055
14	1.079	0.277	0.000	0.332	0.000	0.387	0.001	0.161
15	0.911	0.411	0.019	0.395	0.013	0.379	0.007	0.174
16	1.079	0.276	0.000	0.332	0.000	0.387	0.000	0.160
17	0.915	0.371	0.000	0.381	0.000	0.391	0.000	0.173
18	1.079	0.276	0.000	0.332	0.000	0.387	0.000	0.160
19	0.915	0.371	0.000	0.381	0.000	0.391	0.000	0.173
20	0.909	0.449	0.019	0.445	0.014	0.441	0.009	0.183

TABLE 117.—Uracil.

Atom	P_{rr}	SDN_r	FOD_r	SDR_r	FRD_r	SDE_r	FED_r	π_{rr}
1	1.212	0.590	0.339	1.124	0.619	1.658	0.899	0.426
2	0.853	1.135	1.001	0.925	0.598	0.714	0.195	0.412
3	1.680	0.257	0.157	0.915	0.367	1.576	0.577	0.209
4	0.808	0.699	0.021	0.489	0.016	0.279	0.011	0.204
5	1.750	0.167	0.077	0.826	0.040	1.485	0.003	0.153
6	0.829	0.714	0.206	0.504	0.111	0.293	0.016	0.206
7	1.432	0.394	0.180	0.643	0.180	0.892	0.180	0.232
8	1.437	0.339	0.019	0.588	0.069	0.837	0.119	0.217

TABLE 118(P).—Urobilin.

Atom	P_{rr}	SDN_r	FOD_r	SDR_r	FRD_r	SDE_r	FED_r	π_{rr}
1	0.900	1.480	0.244	1.240	0.283	1.000	0.321	0.496
2	1.092	0.662	0.030	0.912	0.081	1.161	0.131	0.380
3	0.978	1.444	0.281	1.218	0.192	0.991	0.102	0.455
4	1.055	0.619	0.004	0.850	0.170	1.082	0.336	0.353
5	0.789	2.479	0.676	1.677	0.338	0.605	0.000	0.462
6	1.081	0.617	0.027	0.848	0.182	1.079	0.337	0.338
7	1.054	1.143	0.195	1.205	0.162	1.268	0.129	0.461
8	1.106	0.735	0.065	1.010	0.089	1.286	0.113	0.398
9	0.998	1.081	0.151	1.101	0.248	1.122	0.345	0.430
10	0.953	0.487	0.001	0.487	0.002	0.487	0.003	0.239
11	1.040	0.581	0.030	0.601	0.039	0.622	0.048	0.267
12	0.952	0.487	0.001	0.487	0.002	0.487	0.003	0.239
13	1.050	0.534	0.019	0.587	0.035	0.639	0.051	0.265
14	0.951	0.488	0.000	0.488	0.001	0.488	0.001	0.239
15	1.059	0.485	0.004	0.566	0.012	0.646	0.019	0.262
16	0.952	0.487	0.001	0.487	0.001	0.487	0.001	0.239
17	1.046	0.577	0.035	0.600	0.025	0.622	0.015	0.266
18	0.952	0.487	0.001	0.487	0.001	0.487	0.001	0.239
19	1.055	0.541	0.024	0.599	0.022	0.656	0.019	0.266
20	0.951	0.487	0.000	0.487	0.001	0.487	0.001	0.239
21	1.060	0.493	0.008	0.577	0.012	0.661	0.017	0.263
22	1.582	0.584	0.093	0.923	0.047	1.222	0.001	0.277
23	1.334	0.776	0.113	1.092	0.058	1.407	0.003	0.396

TABLE 118(S).—Urobilin.

Atom	P_{rr}	SDN_r	FOD_r	SDR_r	FRD_r	SDE_r	FED_r	π_{rr}
1	0.898	1.432	0.257	1.228	0.295	1.024	0.334	0.473
2	1.099	0.638	0.031	0.914	0.084	1.191	0.137	0.379
3	0.977	1.384	0.290	1.195	0.198	1.006	0.105	0.458
4	1.064	0.606	0.006	0.858	0.178	1.111	0.351	0.352
5	0.801	2.576	0.689	1.591	0.245	0.606	0.000	0.459
6	1.088	0.602	0.030	0.855	0.190	1.107	0.351	0.339
7	1.055	1.098	0.202	1.198	0.168	1.298	0.134	0.464
8	1.113	0.701	0.063	1.011	0.091	1.322	0.118	0.396
9	0.997	1.058	0.162	1.014	0.261	1.149	0.360	0.435
10	1.080	0.276	0.000	0.331	0.002	0.387	0.003	0.161
11	0.908	0.410	0.018	0.396	0.022	0.381	0.027	0.173
12	1.080	0.277	0.000	0.332	0.002	0.387	0.003	0.161
13	0.915	0.385	0.011	0.388	0.020	0.391	0.029	0.173
14	1.080	0.277	0.000	0.333	0.001	0.389	0.001	0.161
15	0.921	0.356	0.002	0.376	0.007	0.395	0.011	0.172
16	1.079	0.277	0.000	0.332	0.001	0.388	0.001	0.161
17	0.913	0.380	0.020	0.394	0.014	0.407	0.008	0.173
18	1.079	0.277	0.000	0.332	0.001	0.388	0.001	0.161
19	0.918	0.387	0.014	0.395	0.012	0.402	0.011	0.173
20	1.079	0.277	0.000	0.333	0.001	0.389	0.001	0.161
21	0.922	0.360	0.005	0.382	0.007	0.404	0.009	0.173
22	1.589	0.533	0.089	0.882	0.045	1.231	0.001	0.269
23	1.346	0.713	0.108	1.074	0.056	1.434	0.004	0.387

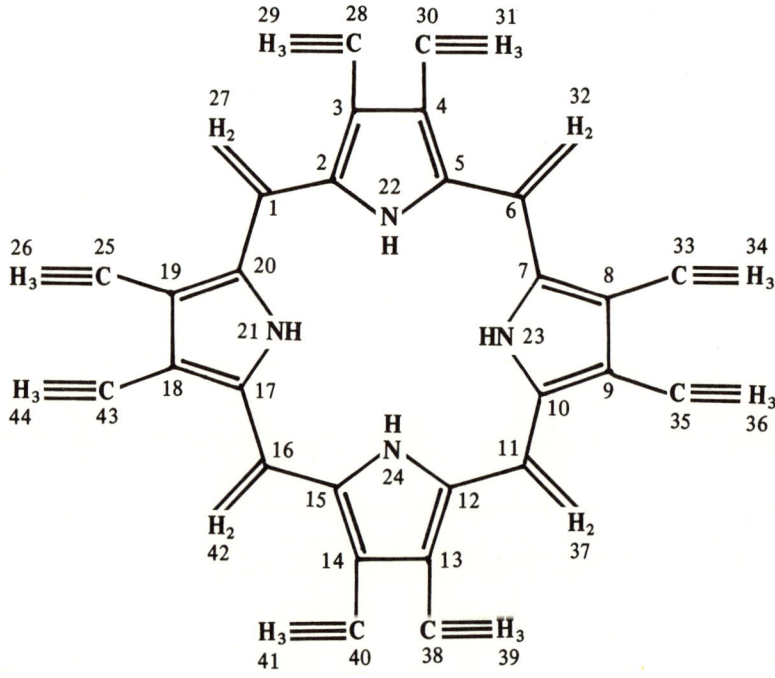

TABLE 119(P).—Uroporphyrin.

Atom	P_{rr}	SDN_r	FOD_r	SDR_r	FRD_r	SDE_r	FED_r	π_{rr}
1	0.955	0.475	0.025	0.475	0.014	0.475	0.004	0.230
2	1.055	0.693	0.115	1.134	0.129	1.576	0.143	0.434
3	1.126	0.547	0.025	0.941	0.042	1.334	0.057	0.371
4	1.126	0.547	0.025	0.941	0.042	1.334	0.057	0.371
5	1.055	0.693	0.115	1.134	0.129	1.576	0.143	0.434
6	0.955	0.475	0.025	0.475	0.014	0.475	0.004	0.230
7	1.055	0.693	0.115	1.134	0.129	1.576	0.143	0.434
8	1.126	0.547	0.025	0.941	0.042	1.334	0.057	0.371
9	1.126	0.547	0.025	0.941	0.042	1.334	0.057	0.371
10	1.055	0.693	0.115	1.134	0.129	1.576	0.143	0.434
11	0.955	0.475	0.025	0.475	0.014	0.475	0.004	0.230
12	1.055	0.693	0.115	1.134	0.129	1.576	0.143	0.434
13	1.126	0.547	0.025	0.941	0.042	1.334	0.057	0.371
14	1.126	0.547	0.025	0.941	0.042	1.334	0.057	0.371
15	1.055	0.693	0.115	1.134	0.129	1.576	0.143	0.434
16	0.955	0.475	0.025	0.475	0.014	0.475	0.004	0.230
17	1.055	0.693	0.115	1.134	0.129	1.576	0.143	0.434
18	1.126	0.547	0.025	0.941	0.042	1.334	0.057	0.371
19	1.126	0.547	0.025	0.941	0.042	1.334	0.057	0.371
20	1.055	0.693	0.115	1.134	0.129	1.576	0.143	0.434
21	1.589	0.347	0.089	0.775	0.045	1.203	0.000	0.247
22	1.589	0.347	0.089	0.775	0.045	1.203	0.000	0.247
23	1.589	0.347	0.089	0.775	0.045	1.203	0.000	0.247
24	1.589	0.347	0.089	0.775	0.045	1.203	0.000	0.247
25	0.951	0.488	0.001	0.488	0.001	0.488	0.000	0.239
26	1.063	0.471	0.005	0.569	0.007	0.667	0.008	0.262
27	1.066	0.517	0.092	0.675	0.086	0.834	0.080	0.281
28	0.951	0.488	0.001	0.488	0.001	0.488	0.000	0.239
29	1.063	0.471	0.005	0.569	0.007	0.667	0.008	0.262
30	0.951	0.488	0.001	0.488	0.001	0.488	0.000	0.239
31	1.063	0.471	0.005	0.569	0.007	0.667	0.008	0.262
32	1.066	0.517	0.092	0.675	0.086	0.834	0.080	0.281
33	0.951	0.488	0.001	0.488	0.001	0.488	0.000	0.239
34	1.063	0.471	0.005	0.569	0.007	0.667	0.008	0.262
35	0.951	0.488	0.001	0.488	0.001	0.488	0.000	0.239
36	1.063	0.471	0.005	0.569	0.007	0.667	0.008	0.262
37	1.066	0.517	0.092	0.675	0.086	0.834	0.080	0.281
38	0.951	0.488	0.001	0.488	0.001	0.488	0.000	0.239
39	1.063	0.471	0.005	0.569	0.007	0.667	0.008	0.262
40	0.951	0.488	0.001	0.488	0.001	0.488	0.000	0.239
41	1.063	0.471	0.005	0.569	0.007	0.667	0.008	0.262
42	1.066	0.517	0.092	0.675	0.086	0.834	0.080	0.281
43	0.951	0.488	0.001	0.488	0.001	0.488	0.000	0.239
44	1.063	0.471	0.005	0.569	0.007	0.667	0.008	0.262

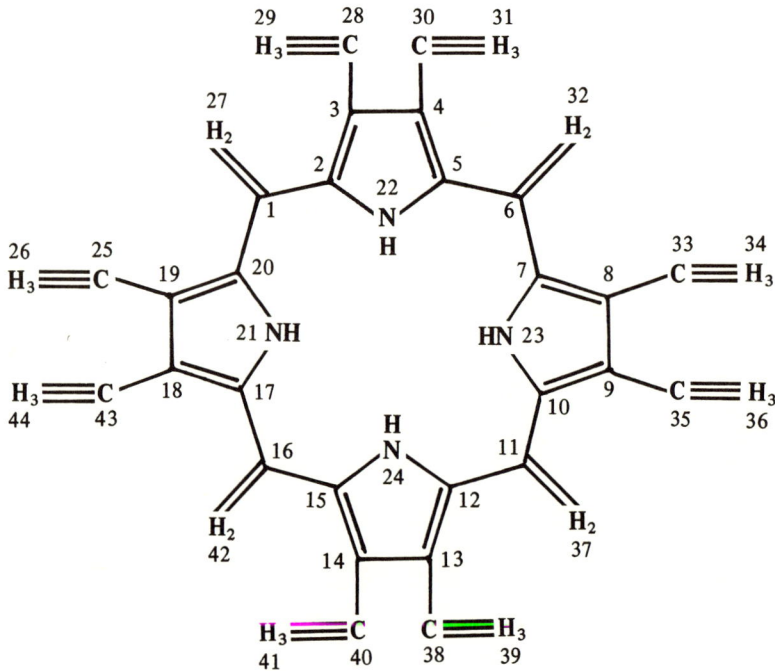

TABLE 119(S).—Uroporphyrin.

Atom	P_{rr}	SDN_r	FOD_r	SDR_r	FRD_r	SDE_r	FED_r	π_{rr}
1	1.077	0.271	0.003	0.329	0.001	0.386	0.000	0.156
2	1.065	0.671	0.143	1.171	0.160	1.671	0.170	0.436
3	1.134	0.528	0.031	0.958	0.046	1.387	0.060	0.369
4	1.134	0.528	0.031	0.958	0.046	1.387	0.060	0.369
5	1.065	0.671	0.143	1.171	0.160	1.671	0.170	0.436
6	1.077	0.271	0.003	0.329	0.001	0.386	0.000	0.156
7	1.065	0.671	0.143	1.171	0.160	1.671	0.170	0.436
8	1.134	0.528	0.031	0.958	0.046	1.387	0.067	0.369
9	1.134	0.528	0.031	0.958	0.046	1.387	0.067	0.369
10	1.065	0.671	0.143	1.171	0.160	1.671	0.170	0.436
11	1.077	0.271	0.003	0.329	0.001	0.386	0.000	0.156
12	1.065	0.671	0.143	1.171	0.160	1.671	0.170	0.436
13	1.134	0.528	0.031	0.958	0.046	1.387	0.067	0.369
14	1.134	0.528	0.031	0.958	0.046	1.387	0.067	0.369
15	1.065	0.671	0.143	1.171	0.160	1.671	0.170	0.436
16	1.077	0.271	0.003	0.329	0.001	0.386	0.000	0.156
17	1.065	0.671	0.143	1.171	0.160	1.671	0.170	0.436
18	1.134	0.528	0.031	0.958	0.046	1.387	0.067	0.369
19	1.134	0.528	0.031	0.958	0.046	1.387	0.067	0.369
20	1.065	0.671	0.143	1.171	0.160	1.671	0.170	0.436
21	1.596	0.325	0.096	0.769	0.048	1.213	0.000	0.240
22	1.596	0.325	0.096	0.769	0.048	1.213	0.000	0.240
23	1.596	0.325	0.096	0.769	0.048	1.213	0.000	0.240
24	1.696	0.325	0.096	0.769	0.048	1.213	0.000	0.240
25	1.080	0.277	0.000	0.333	0.000	0.389	0.000	0.161
26	0.924	0.348	0.003	0.379	0.004	0.409	0.005	0.172
27	0.922	0.381	0.049	0.453	0.050	0.526	0.000	0.182
28	1.080	0.277	0.000	0.333	0.000	0.389	0.000	0.161
29	0.924	0.348	0.003	0.379	0.004	0.409	0.005	0.172
30	1.080	0.277	0.000	0.333	0.000	0.389	0.000	0.161
31	0.924	0.348	0.003	0.379	0.004	0.409	0.005	0.172
32	0.922	0.381	0.049	0.453	0.050	0.526	0.000	0.182
33	1.080	0.277	0.000	0.333	0.000	0.389	0.000	0.161
34	0.924	0.348	0.003	0.379	0.004	0.409	0.005	0.172
35	1.080	0.277	0.000	0.333	0.000	0.389	0.000	0.161
36	0.924	0.348	0.003	0.379	0.004	0.409	0.005	0.172
37	0.922	0.381	0.049	0.453	0.050	0.526	0.000	0.182
38	1.080	0.277	0.000	0.333	0.000	0.389	0.000	0.161
39	0.924	0.348	0.003	0.379	0.004	0.409	0.005	0.172
40	1.080	0.277	0.000	0.333	0.000	0.389	0.000	0.161
41	0.924	0.348	0.003	0.279	0.004	0.409	0.005	0.172
42	0.922	0.381	0.049	0.453	0.050	0.526	0.000	0.182
43	1.080	0.277	0.000	0.333	0.000	0.389	0.000	0.161
44	0.924	0.348	0.003	0.379	0.004	0.409	0.005	0.172

TABLE 120(P).—Vitamin E (Tocopherol).

Atom	P_{rr}	SDN_r	FOD_r	SDR_r	FRD_r	SDE_r	FED_r	π_{rr}
1	0.999	0.731	0.000	0.895	0.232	1.058	0.463	0.382
2	1.030	0.734	0.408	0.893	0.280	1.051	0.151	0.396
3	1.030	0.734	0.408	0.893	0.280	1.051	0.151	0.396
4	0.999	0.731	0.000	0.895	0.232	1.058	0.463	0.382
5	1.030	0.734	0.408	0.893	0.280	1.051	0.151	0.396
6	1.030	0.734	0.408	0.893	0.280	1.051	0.151	0.396
7	1.930	0.054	0.000	0.587	0.092	1.120	0.184	0.041
8	1.930	0.054	0.000	0.587	0.092	1.120	0.184	0.041
9	0.952	0.488	0.017	0.488	0.010	0.488	0.002	0.239
10	1.053	0.493	0.075	0.563	0.049	0.632	0.023	0.263
11	0.952	0.488	0.017	0.488	0.010	0.488	0.002	0.239
12	1.053	0.493	0.075	0.563	0.049	0.632	0.023	0.263
13	0.952	0.488	0.017	0.488	0.010	0.488	0.002	0.239
14	1.053	0.493	0.075	0.567	0.049	0.632	0.023	0.263
15	0.952	0.488	0.017	0.488	0.010	0.488	0.002	0.239
16	1.053	0.493	0.075	0.563	0.049	0.632	0.023	0.263

TABLE 120(S).—Vitamin E (Tocopherol).

Atom	P_{rr}	SDN_r	FOD_r	SDR_r	FRD_r	SDE_r	FED_r	π_{rr}
1	1.010	0.708	0.000	0.893	0.237	1.078	0.475	0.380
2	1.033	0.716	0.461	0.894	0.308	1.072	0.154	0.396
3	1.033	0.716	0.461	0.894	0.308	1.072	0.154	0.396
4	1.010	0.708	0.000	0.893	0.237	1.078	0.475	0.380
5	1.033	0.716	0.461	0.894	0.308	1.072	0.154	0.396
6	1.033	0.716	0.461	0.894	0.308	1.072	0.154	0.396
7	1.932	0.051	0.000	0.589	0.094	1.126	0.189	0.040
8	1.932	0.051	0.000	0.589	0.094	1.126	0.189	0.040
9	1.080	0.276	0.001	0.332	0.001	0.388	0.002	0.161
10	0.917	0.361	0.038	0.374	0.025	0.386	0.013	0.172
11	1.080	0.276	0.001	0.332	0.001	0.388	0.002	0.161
12	0.917	0.361	0.038	0.374	0.025	0.386	0.013	0.172
13	1.080	0.276	0.001	0.332	0.001	0.388	0.002	0.161
14	0.917	0.361	0.038	0.374	0.025	0.386	0.013	0.172
15	1.080	0.276	0.001	0.332	0.001	0.388	0.002	0.161
16	0.917	0.361	0.038	0.374	0.025	0.386	0.013	0.172

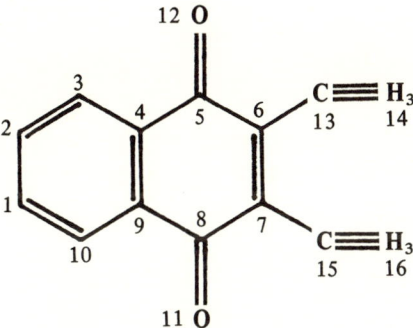

TABLE 121(P).—Vitamin K.

Atom	P_π	SDN_r	FOD_r	SDR_r	FRD_r	SDE_r	FED_r	π_π
1	0.979	0.997	0.071	0.898	0.048	0.799	0.025	0.405
2	0.979	0.997	0.071	0.898	0.048	0.799	0.025	0.405
3	0.968	0.997	0.034	0.898	0.017	0.799	0.001	0.414
4	0.999	0.976	0.106	0.878	0.069	0.779	0.033	0.372
5	0.820	1.063	0.166	0.668	0.084	0.272	0.003	0.206
6	0.945	1.468	0.296	1.199	0.490	0.930	0.684	0.471
7	0.945	1.468	0.296	1.199	0.490	0.930	0.684	0.471
8	0.820	1.063	0.166	0.688	0.084	0.272	0.003	0.206
9	0.999	0.976	0.106	0.878	0.069	0.779	0.033	0.372
10	0.968	0.997	0.034	0.898	0.017	0.799	0.001	0.414
11	1.294	1.168	0.290	0.903	0.192	0.638	0.093	0.285
12	1.294	1.168	0.290	0.903	0.192	0.638	0.093	0.285
13	0.952	0.486	0.001	0.486	0.012	0.486	0.024	0.239
14	1.042	0.579	0.037	0.596	0.087	0.613	0.137	0.267
15	0.952	0.486	0.001	0.486	0.012	0.486	0.024	0.239
16	1.042	0.579	0.037	0.596	0.087	0.613	0.137	0.267

TABLE 121(S).—Vitamin K.

Atom	P_π	SDN_r	FOD_r	SDR_r	FRD_r	SDE_r	FED_r	π_π
1	0.979	0.994	0.078	0.896	0.054	0.799	0.031	0.405
2	0.979	0.994	0.078	0.896	0.054	0.799	0.031	0.405
3	0.968	0.994	0.035	0.896	0.018	0.799	0.001	0.414
4	1.000	0.974	0.116	0.877	0.077	0.799	0.039	0.372
5	0.820	1.052	0.173	0.662	0.088	0.273	0.003	0.207
6	0.948	1.394	0.281	1.171	0.513	0.949	0.744	0.472
7	0.948	1.394	0.281	1.171	0.513	0.949	0.744	0.472
8	0.820	1.052	0.173	0.662	0.088	0.273	0.003	0.207
9	1.000	0.974	0.116	0.877	0.077	0.799	0.039	0.372
10	0.968	0.994	0.035	0.896	0.018	0.799	0.001	0.414
11	1.295	1.139	0.297	0.890	0.198	0.641	0.099	0.285
12	1.295	1.139	0.297	0.890	0.198	0.641	0.099	0.285
13	1.079	0.276	0.000	0.332	0.007	0.387	0.014	0.161
14	0.911	0.407	0.020	0.392	0.045	0.376	0.070	0.173
15	1.079	0.276	0.000	0.332	0.007	0.387	0.014	0.161
16	0.911	0.407	0.020	0.392	0.045	0.376	0.070	0.173

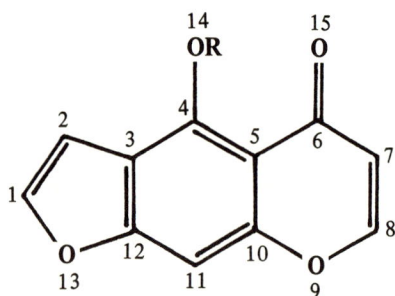

TABLE 122.—Visnagin.

Atom	P_{rr}	SDN_r	FOD_r	SDR_r	FRD_r	SDE_r	FED_r	π_{rr}
1	0.994	0.924	0.107	1.141	0.253	1.358	0.399	0.493
2	1.114	0.651	0.053	0.955	0.159	1.259	0.265	0.402
3	1.093	0.562	0.023	0.779	0.093	0.996	0.163	0.329
4	0.915	0.970	0.361	0.977	0.311	0.983	0.261	0.407
5	1.118	0.590	0.018	0.820	0.013	1.050	0.008	0.354
6	0.837	0.770	0.221	0.531	0.111	0.291	0.002	0.207
7	1.150	0.674	0.133	1.008	0.118	1.342	0.103	0.430
8	0.838	1.236	0.562	0.997	0.286	0.757	0.010	0.445
9	1.801	0.198	0.055	0.610	0.075	1.023	0.004	0.114
10	0.940	0.836	0.001	0.863	0.093	0.890	0.185	0.374
11	1.126	0.672	0.135	1.021	0.222	1.370	0.310	0.419
12	0.968	0.798	0.065	0.826	0.049	0.852	0.033	0.373
13	1.793	0.187	0.001	0.563	0.034	0.938	0.068	0.113
14	1.914	0.086	0.038	0.587	0.063	1.088	0.087	0.054
15	1.375	0.510	0.228	0.678	0.120	0.846	0.012	0.254

TABLE 123.—Xanthine.

Atom	P_{rr}	SDN_r	FOD_r	SDR_r	FRD_r	SDE_r	FED_r	π_{rr}
1	1.726	0.172	0.032	1.037	0.228	1.003	0.425	0.166
2	0.810	0.694	0.008	0.494	0.011	0.293	0.013	0.205
3	1.755	0.151	0.044	0.822	0.022	1.493	0.000	0.147
4	0.823	0.723	0.356	0.513	0.186	0.303	0.015	0.207
5	1.187	0.467	0.144	1.031	0.289	1.594	0.434	0.342
6	1.287	0.553	0.493	0.890	0.331	1.227	0.169	0.360
7	1.002	0.825	0.522	1.122	0.478	1.419	0.434	0.460
8	1.585	0.337	0.025	0.753	0.026	1.168	0.027	0.244
9	0.955	0.788	0.115	0.942	0.209	1.096	0.302	0.386
10	1.441	0.332	0.006	0.608	0.044	0.885	0.083	0.216
11	1.420	0.394	0.254	0.645	0.175	0.895	0.097	0.232

24 H₃ 26 H₃ 28 H₃
23 C   25 C   27 C

H₃C
22 21

1 3 5 7 9 11 13 15 17 19
   2 4 6 8 10 12 14 16 18 20

29 C   31 C   C 33
30 H₃ 32 H₃ H₃ 34

TABLE 124(P).—Xanthophyll.

Atom	P_{rr}	SDN_r	FOD_r	SDR_r	FRD_r	SDE_r	FED_r	π_{rr}
1	0.944	1.934	0.188	2.264	0.174	2.953	0.159	0.681
2	1.031	0.834	0.003	0.937	0.005	1.041	0.008	0.409
3	0.992	1.663	0.197	1.993	0.166	2.323	0.153	0.553
4	1.024	1.034	0.018	1.137	0.017	1.241	0.015	0.454
5	0.960	1.488	0.158	1.818	0.151	2.148	0.144	0.493
6	1.031	1.102	0.032	1.315	0.036	1.528	0.041	0.473
7	1.002	1.395	0.133	1.724	0.131	2.054	0.130	0.497
8	1.027	1.203	0.061	1.417	0.058	1.630	0.056	0.483
9	0.971	1.281	0.102	1.611	0.108	1.941	0.114	0.466
10	1.028	1.235	0.079	1.565	0.086	1.894	0.094	0.491
11	1.016	1.215	0.071	1.544	0.083	1.474	0.095	0.482
12	1.012	1.315	0.109	1.644	0.111	1.974	0.113	0.492
13	1.032	1.133	0.042	1.463	0.059	1.793	0.077	0.482
14	0.966	1.387	0.136	1.717	0.133	2.047	0.131	0.478
15	1.031	1.088	0.027	1.301	0.034	1.514	0.041	0.471
16	0.995	1.519	0.159	1.849	0.152	2.178	0.145	0.516
17	1.039	0.958	0.008	1.171	0.018	1.384	0.027	0.453
18	0.947	1.656	0.173	1.986	0.165	2.315	0.158	0.533
19	1.038	0.804	0.002	0.907	0.005	1.010	0.007	0.402
20	0.930	1.979	0.180	2.309	0.172	2.609	0.164	0.718
21	0.953	0.484	0.000	0.484	0.000	0.484	0.000	0.238
22	1.043	0.631	0.023	0.722	0.021	0.812	0.010	0.273
23	0.953	0.484	0.000	0.484	0.000	0.484	0.000	0.238
24	1.043	0.631	0.023	0.722	0.021	0.812	0.010	0.273
25	0.953	0.487	0.000	0.487	0.000	0.487	0.000	0.239
26	1.046	0.583	0.019	0.673	0.019	0.764	0.018	0.267
27	0.953	0.487	0.000	0.487	0.000	0.487	0.000	0.239
28	1.047	0.558	0.012	0.648	0.013	0.739	0.014	0.266
29	0.953	0.487	0.000	0.487	0.000	0.487	0.000	0.239
30	1.046	0.551	0.017	0.661	0.016	0.752	0.016	0.266
31	0.953	0.486	0.000	0.486	0.000	0.486	0.000	0.239
32	1.044	0.602	0.021	0.693	0.020	0.783	0.020	0.268
33	0.953	0.484	0.000	0.484	0.000	0.484	0.000	0.238
34	1.042	0.635	0.022	0.725	0.021	0.816	0.020	0.274

TABLE 124(S).—Xanthophyll.

Atom	P_π	SDN_r	FOD_r	SDR_r	FRD_r	SDE_r	FED_r	π_π
1	0.910	1.303	0.199	2.399	0.178	2.894	0.157	0.696
2	1.053	0.755	0.001	0.941	0.006	1.126	0.011	0.401
3	0.983	1.617	0.193	2.113	0.172	2.609	0.150	0.557
4	1.038	0.961	0.016	1.147	0.017	1.332	0.018	0.449
5	0.950	1.442	0.171	1.938	0.156	2.343	0.142	0.498
6	1.047	1.011	0.027	1.364	0.037	1.718	0.047	0.468
7	1.001	1.341	0.143	1.837	0.136	2.333	0.130	0.499
8	1.039	1.117	0.059	1.470	0.060	1.823	0.060	0.479
9	0.966	1.228	0.107	1.724	0.111	2.220	0.116	0.471
10	1.042	1.136	0.075	1.675	0.089	2.214	0.103	0.488
11	1.022	1.151	0.072	1.646	0.086	2.142	0.099	0.482
12	1.020	1.224	0.109	1.764	0.114	0.303	0.119	0.492
13	1.044	1.060	0.040	1.556	0.061	2.052	0.083	0.479
14	0.963	1.308	0.139	1.847	0.137	2.386	0.135	0.482
15	1.042	1.028	0.027	1.339	0.035	1.650	0.043	0.467
16	0.995	1.440	0.166	1.979	0.157	2.618	0.148	0.518
17	1.054	0.894	0.006	1.205	0.019	1.516	0.031	0.448
18	0.937	1.584	0.180	2.123	0.169	2.663	0.159	0.538
19	1.052	0.761	0.002	0.906	0.005	1.051	0.008	0.397
20	0.912	1.912	0.187	2.451	0.176	2.991	0.164	0.727
21	1.079	0.276	0.000	0.333	0.000	0.389	0.000	0.160
22	0.908	0.442	0.014	0.477	0.013	0.512	0.011	0.176
23	1.079	0.276	0.000	0.333	0.000	0.389	0.000	0.160
24	0.908	0.442	0.014	0.447	0.013	0.512	0.011	0.176
25	1.080	0.277	0.000	0.333	0.000	0.390	0.000	0.160
26	0.911	0.411	0.012	0.447	0.011	0.482	0.010	0.173
27	1.080	0.277	0.000	0.333	0.000	0.390	0.000	0.160
28	0.913	0.396	0.008	0.432	0.008	0.467	0.009	0.173
29	1.080	0.277	0.000	0.333	0.000	0.390	0.000	0.161
30	0.912	0.402	0.010	0.440	0.010	0.479	0.010	0.173
31	1.079	0.277	0.000	0.333	0.000	0.390	0.000	0.161
32	0.911	0.421	0.013	0.459	0.012	0.498	0.011	0.174
33	1.079	0.276	0.000	0.333	0.000	0.389	0.000	0.160
34	0.908	0.442	0.013	0.480	0.012	0.519	0.012	0.176

TABLE 125.—Xanthopterin.

Atom	P_{π}	SDN_r	FOD_r	SDR_r	FRD_r	SDE_r	FED_r	π_{rr}
1	0.851	0.699	0.052	0.564	0.046	0.428	0.041	0.217
2	0.961	1.349	0.615	1.354	0.398	1.360	0.181	0.540
3	1.123	1.099	0.428	0.963	0.218	0.828	0.009	0.415
4	0.993	0.705	0.028	0.928	0.099	1.151	0.170	0.349
5	1.417	0.418	0.024	1.332	0.189	2.245	0.353	0.343
6	0.812	0.974	0.000	0.866	0.053	0.759	0.106	0.337
7	1.675	0.275	0.030	0.855	0.031	1.435	0.031	0.210
8	0.837	0.741	0.077	0.566	0.055	0.391	0.032	0.215
9	1.016	1.008	0.346	1.403	0.356	1.797	0.365	0.479
10	1.633	0.488	0.227	1.168	0.232	1.847	0.236	0.287
11	1.453	0.405	0.068	0.862	0.125	1.319	0.182	0.238
12	1.422	0.463	0.102	0.809	0.122	1.156	0.141	0.246
13	1.806	0.165	0.000	1.078	0.077	1.991	0.153	0.152

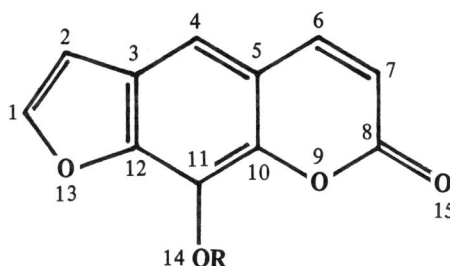

TABLE 126.—Xanthotoxin (8-methoxypsoralen).

Atom	P_{rr}	SDN_r	FOD_r	SDR_r	FRD_r	SDE_r	FED_r	π_{rr}
1	0.971	0.973	0.041	1.095	0.165	1.217	0.289	0.488
2	1.123	0.646	0.028	0.957	0.150	1.267	0.272	0.402
3	1.053	0.612	0.011	0.734	0.048	0.856	0.086	0.326
4	1.040	1.007	0.337	1.183	0.370	1.359	0.402	0.497
5	1.049	0.629	0.057	0.751	0.029	0.873	0.000	0.331
6	0.991	1.235	0.517	0.990	0.260	0.746	0.003	0.426
7	1.071	1.010	0.440	1.132	0.221	1.254	0.001	0.493
8	0.811	0.777	0.107	0.532	0.055	0.287	0.002	0.207
9	1.838	0.157	0.045	0.585	0.071	1.014	0.097	0.091
10	0.990	0.840	0.078	0.997	0.205	1.153	0.331	0.412
11	1.018	0.748	0.070	0.954	0.188	1.161	0.306	0.394
12	1.009	0.796	0.125	0.901	0.063	1.007	0.001	0.393
13	1.793	0.192	0.003	0.576	0.047	0.960	0.091	0.114
14	1.931	0.059	0.008	0.601	0.057	1.143	0.105	0.041
15	1.393	0.495	0.134	0.649	0.074	0.803	0.014	0.242

References

(1) K. Fukui, T. Yonezawa, C. Nagata, and H. Shingu, *J. Chem. Phys.* **22**, 1433 (1954).

(2) K. Fukui, T. Yonezawa, and H. Shingu, *J. Chem. Phys.*, **20**, 722 (1952).

(3) K. Fukui, T. Yonezawa, and C. Nagata, *J. Chem. Phys.*, **27**, 1247 (1957).

(4) C. A. Coulson and H. C. Longuet-Higgins, *Proc. Roy. Soc.* (London), **A191**, 39; **A192**, 16 (1947).

(5) B. Pullman and A. Pullman, *Quantum Biochemistry*, Interscience, New York, pp 104-110, 541-545 (1965).

(6) A. Streitwieser, Jr., *Molecular Orbital Theory for Organic Chemists*, John Wiley and Sons, New York, pp 131-135 (1961).

(7) R. S. Mulliken, G. A. Rieke, and W. G. Brown, *J. Am. Chem. Soc.*, **63**, 41 (1941).

(8) G. W. Wheland and L. Pauling, *J. Am. Chem. Soc.*, **57**, 2086 (1953).

(9) H. C. Longuet-Higgins *J. Chem. Phys.*, **18**, 283 (1950).

(10) M. Sun and P. S. Song, *Biochemistry* **12**, 4663 (1973).

(11) P. S. Song and M. Sun, pp 407-429, in *Chemical and Biochemical Reactivity: Jerusalem Symp. on Quantum Chemistry and Biochemistry*, vol. VI., Israel Academy of Sciences, Jerusalem (1974).

APPENDIX A.—*Definitions of reactivity indices.*

Reactivity Index	Definition (in equation)	Explanation
π-electron density	$P_{rr} = 2 \sum\limits_{j}^{OCC} c_{ri}^2$	Where c_r is LCAO expansion coefficients obtained from HMO calculation.
Superdelocalizablity for nucleophilic attack	$SDN_r = 2 \sum\limits_{j}^{unocc} \dfrac{(c_r^j)^2}{-k_j}$	According to frontier electron theory SDN_r is a measure of the stabilization energy of its transition state complex at the rth atom, where the sum is overall unoccupied molecular orbitals and k_j is the MO energy levels.
Frontier orbital density	$FOD_r = 2(c_r^{LVMO})^2$	According to frontier electron theory FOD_r is a measure of the reactivity of nucleophilic attack at position r. The larger the value of frontier orbital density, the more reactive is the nucleophilic attack for the position, because the accommodation of a nucleophile by the substrate is dependent upon the availability of the LVMO at the position of reaction.
Superdelocalizability for radical attack	$SRD_r = \sum\limits_{j}^{occ} \dfrac{(c_r^j)^2}{k_j} + \sum\limits_{j}^{unocc} \dfrac{(c_r^j)^2}{-k_j}$	SRD_r is a measure of the stabilization energy of the transition state complex at the rth atom.
Frontier Radical density	$FRD_r = (c_r^{HOMO})^2 + (c_r^{LVMO})^2$	Frontier radical density is a measure of the reactivity of atom r toward a radical reagent.
Superdelocalizability for electrophilic attack	$SDE_r = 2 \sum\limits_{j}^{occ} \dfrac{(c_r^j)^2}{k_j}$	Superdelocalizability for electrophilic attack is a measure of the stabilization energy of its transition state complex at the rth atom.
Frontier electron density	$FED_r = 2(c_r^{HOMO})^2$	According to frontier electron theory the frontier electron density is a measure of the reactivity of electrophilic attack at the position r.
Atom-atom polarizability	$\pi_{rr} = \left(\dfrac{\partial P_{rr}}{\partial \alpha_{rr}}\right) = \dfrac{1}{\beta} \sum\limits_{i}^{occ} \sum\limits_{j}^{unocc} n_i(2-n_i)\left\{\dfrac{(c_r^i)^2(c_r^j)^2}{k_i - k_j}\right\}$ and $\partial E_\pi = P_{rr}\partial \alpha_r + \dfrac{1}{2}\pi_{rr}(\partial \alpha_r)^2 + \cdots$	Where n_i and n_j are the number of π-electrons occupying ith and jth MO's, respectively, and ∂E_π is the change in total π- electron energy with respect to the change in rth Coulomb integral (α_r).

APPENDIX B.—*Coulomb (α) and resonance (β) integral parameters.*

Parameter		C=N−	C=$\overset{+}{\text{N}}$H−	C−N−	C=O	C−O−	C−O·	C−Cl	C−F	C−I	C−H	C−S−	C=S	C−C	C≡CH₃
Pullman (P)	α	0.4	2.0	1.0	0.2 1.2	2.0	1.2	0.18 1.80	0.2 3.0	0.12 1.30	−1.8	0.9		−0.1	−0.2
	β	1.0	1.0	0.9	2.0	0.6	0.9	0.8	0.7	0.3	0.2	0.5	1.2	0.7	2.0
Streitwieser (S)	α	0.4	2.0	1.0	0.2 1.2	2.0	1.2	0.18 1.80	0.2 3.0	0.12 1.30	−1.8	0.9		−0.1	−0.5
	β	1.0	1.0	0.9	2.0	0.6	0.9	0.8	0.7	0.3	0.2	0.5	1.2	0.8	3.0
Fukui (F)	α	0.1 0.6			0.2 2.0	0.6									
	β	1.0			1.414	0.7									